# Process Development: Physicochemical Concepts

**John H. Atherton**

Business Research Associate, AVECIA Huddersfield

**Keith J. Carpenter**

Company Research Associate, ZENECA Huddersfield Works

OXFORD

UNIVERSITY PRESS

*This book has been printed digitally and produced in a standard specification*
*in order to ensure its continuing availability*

# OXFORD
UNIVERSITY PRESS

Great Clarendon Street, Oxford OX2 6DP

Oxford University Press is a department of the University of Oxford.
It furthers the University's objective of excellence in research, scholarship,
and education by publishing worldwide in

Oxford  New York

Auckland   Cape Town  Dar es Salaam   Hong Kong   Karachi
Kuala Lumpur  Madrid  Melbourne  Mexico City  Nairobi
New Delhi  Shanghai  Taipei  Toronto
With offices in
Argentina  Austria  Brazil  Chile  Czech Republic  France  Greece
Guatemala  Hungary  Italy  Japan  South Korea  Poland  Portugal
Singapore  Switzerland  Thailand  Turkey  Ukraine  Vietnam

Oxford is a registered trade mark of Oxford University Press
in the UK and in certain other countries

Published in the United States
by Oxford University Press Inc., New York

ISBN  0-19-850372-5

Antony Rowe Ltd., Eastbourne

# OXFORD CHEMISTRY PRIMERS

**Physical Chemistry Editor**
RICHARD G. COMPTON
University of Oxford

**Founding/Organic Editor**
STEPHEN G. DAVIES
University of Oxford

**Inorganic Chemistry Editor**
JOHN EVANS
University of Southampton

**Chemical Engineering Editor**
LYNN F. GLADDEN
University of Cambridge

# Series Editor's Foreword

Oxford Chemistry Primers are designed to provide clear and concise introductions to a wide range of topics that may be encountered by chemistry students as they progress from the freshman stage through to graduation. The Physical Chemistry series contains books easily recognised as relating to established fundamental core material that all chemists need to know, as well as books reflecting new directions and research trends in the subject, thereby anticipating (and perhaps encouraging) the evolution of modern undergraduate courses.

In this Physical Chemistry Primer John Atherton and Keith Carpenter present an introductory account of the physicochemical basis of process development. The primer clearly explains and illustrates the basic ideas and applications of a subject which is vital knowledge for any would-be practising industrial chemist. This Primer will be of interest to all students of science (and their mentors).

Richard G. Compton
*Physical and Theoretical Chemistry Laboratory,*
*University of Oxford*

# Preface

Despite the fact that many chemists and chemical engineers who work in the fine chemicals industry are involved in chemical process development, there are very few textbooks dealing with the subject. Two excellent recent texts cover the subjects of route selection and process optimisation from the viewpoint of the organic chemist. This Primer describes a complementary approach with an emphasis on understanding the physicochemical aspects of process optimisation and scale-up. It crosses the boundary between chemistry and chemical engineering, and seeks to show how these disciplines must be integrated to provide a comprehensive approach to the subject.

Two other Primers complement this text, and we commend them to those readers who find this volume useful.

We thank many of our colleagues for constructive criticism and ZENECA for providing such interesting employment and for permission to publish.

*Huddersfield*                                                                        J. H. A. and K. J. C.
February, 1999

# Contents

# 1   The scope of process development

The focus of this book is on the development of manufacturing processes for fine chemicals: those complex organic chemicals required as intermediates or active ingredients for pharmaceuticals, agrochemicals or 'specialty' chemicals. Figs 1.1–1.4 show some examples of the types of compounds which are made. Process development is an essential part of the chain of events preceding the introduction of a new chemical product to the market place. It is the link between the research department, who discover the new compound, and the manufacturing departments, who make the product for sale.

## 1.1   Objectives of process development

Process development has a number of objectives.
* Converting a research synthesis into a viable manufacturing process.
* Providing information to enable process design for robust operation.
* Providing design information for equipment selection and sizing.
* Enabling continuous improvement in the operation of an existing process
* Discovering step changes in technology.

In the discovery of new products the chemists involved will make their compounds by whichever synthetic route is most convenient on the laboratory scale. This may involve the use of several techniques which may not be economic on the manufacturing scale. Examples of such techniques include the use of expensive reagents such as metal hydrides, the use of solvents which are toxic or difficult to recover, chromatographic separations, and multiple solvent extractions. Typically, many thousands of compounds will be synthesised during the discovery phase for each one which reaches the development stage. Design of the synthesis at the discovery stage will not take into account materials costs, yield, equipment capital costs, nor the amount and nature of waste material generated. Direct implementation for manufacture of the 'discovery route' will rarely be possible. Costs of materials and capital cost requirements will probably be much too high to be economic, and the quantity of effluent will probably be unacceptable. It is the job of the process development chemist to chose the most cost effective route for the selected compounds, and to develop the chosen route to give a process which meets a range of requirements. Process development is thus a key part of the product development strategy. It makes the difference between having a synthesis which is commercially viable rather than one which is not. Skilled process development can also result in materials savings, over the lifetime of a large product, in the region of £0.1–1Bn *vs* direct implementation of the processes used at the product discovery stage.

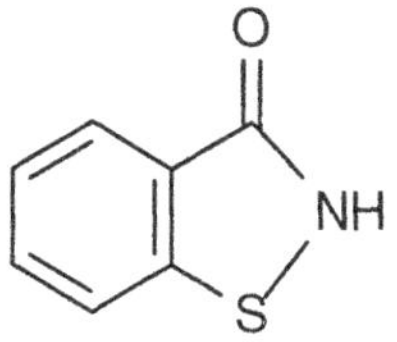

A manufacturing process must meet financial, environmental, health and safety targets in order to be viable. Quality is of primary importance. It is

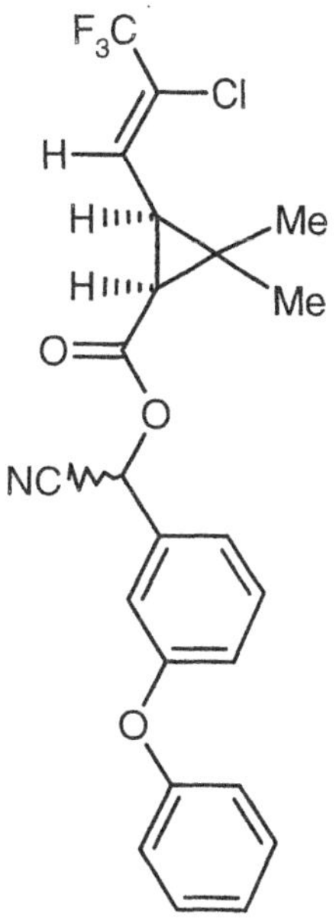

**Fig. 1.1** A synthetic pyrethroid, used as an insecticide.

**Fig. 1.2** Benzisothiazolinone, used as a biocide

Major financial savings result from expert process development.

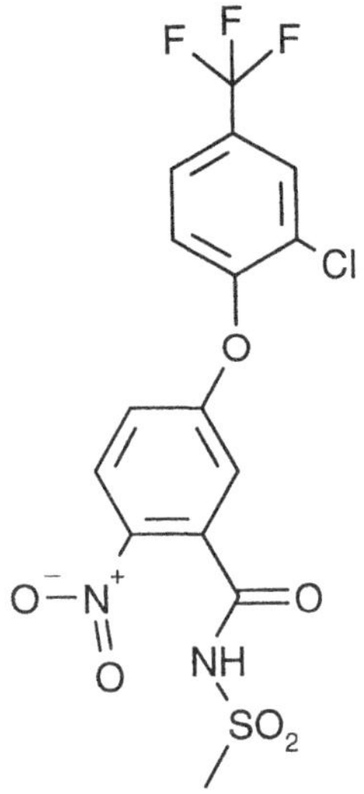

**Fig. 1.3** A substituted diphenyl ether, used as a herbicide.

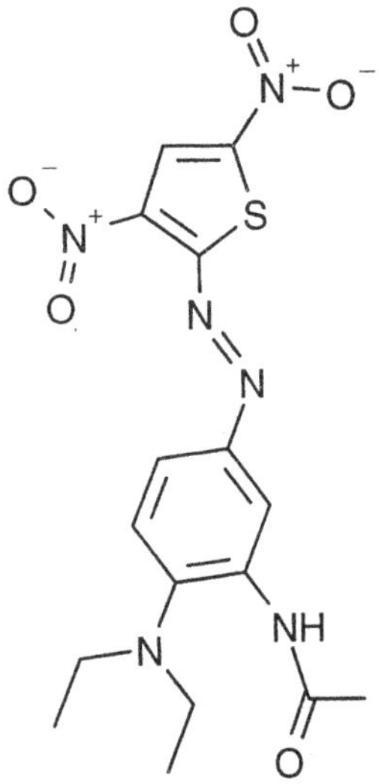

**Fig. 1.4** A green azo dyestuff, used for dyeing polyester.

essential that product meets the specification defined by the user, otherwise it will be unsaleable. The projected materials costs will be based on yields estimated from the laboratory development work. For the product to be a financial success these yields must then be achieved on the full scale. The operating cost and sales volume will be based on a plant design capacity. In order to achieve the design capacity and meet forecast demand, the process must operate in a trouble free manner. It must be capable of routine operation without intervention by skilled chemists. Safety and health targets based on legislation and corporate standards must be met. Process effluent must be kept to a minimum and rendered acceptable for disposal.

Process development for a new product usually starts after a chemical route to a required compound has been selected, but the procedure can be iterative, in that improved process understanding may lead to improved performance and thus influence the relative merits of alternative manufacturing routes.

## 1.2   The approach to process development

In order to be able to optimise processes on the laboratory and manufacturing scale, the scientists involved in process development must have an appreciation of the physico-chemical principles underlying synthesis. The topics discussed here are only part of the total work needed to take a process from laboratory to full scale. The development is an interdisciplinary activity involving a team of chemists and chemical engineers. This team will have very many aspects to consider, including: costs and supply of raw materials; toxicity and safe handling of materials and product; specification of the most cost effective equipment; specification and testing of materials of construction; evaluation of chemical and operational hazards and specification of safety protection systems; minimising the waste material generated, and defining the recycle, safe disposal or destruction of such waste as is produced; design and costing of the process plant; as well as resource planning of people, laboratory equipment and semi-technical scale equipment trials. It is hardly surprising that, faced with this seeming mountain of topics which are not related to the science of the process, finding time to think about the intricacies of the process chemistry can be difficult.

This Primer is intended to provide a framework for thinking about the science underlying process development.

- Chapter 2 discusses the overall strategy for process development.
- Chapters 3 and 4 cover equilibria, rates, and competing reactions in homogeneous systems.
- Chapter 5 introduces the concept of chemical selectivity being coupled to mixing rates in nominally homogeneous systems.
- Chapter 6 describes the quantitative treatment of equilibria in two-phase systems, which is relevant to both the reaction and workup stages of a process.
- In Chapter 7 the principles involved in obtaining liquid-liquid dispersions are discussed, and the concept of mass transfer in two-phase systems is introduced.

- Chapter 8 provides an introduction to the principles governing reaction rates and selectivities in two-phase reaction systems.

- Work-up tends to be a neglected part of process development. Chapter 9 outlines the areas where an appreciation of the science is essential.

- In Chapter 10 an outline is given of the thought processes necessary to ensure secure scale-up from laboratory scale to manufacture.

## 1.3   Statistical design in process development

Statistical techniques are widely used in process development. This involves design of a set, or sets, of experiments in which those variables which are considered to be important are systematically changed. Analysis of the results permits identification of the most important variables and of the optimum processing conditions. Acceptance of the need for the technique admits that the variables contributing to process performance are insufficiently understood to permit a process design from first principles. For the statistical approach to be effective the relevant parameters must be investigated. A knowledge of the physicochemical principles underlying the process is of great help in selecting the variables to be evaluated in a statistically designed set of experiments.

## 1.4   A 'total technology' output

The output from the process development exercise is a *total process*. It includes handling feed materials, reaction stages, purifications, recycles, treatment of waste streams, and possibly product formulation, packaging and storage. The focus of this Primer is on the chemical transformations and product isolations which are at the heart of process performance and where solutions to many of the problems require constant interaction between chemical engineers and chemists. It is intended to contribute to bridging the knowledge gap between the synthetic chemists and the chemical engineers involved in process development. If it helps them to understand each others priorities and expertise, and to see the solution to a problem as a shared objective, it will have gone a long way to achieving its goals.

# 2 Strategy for process development

In this Chapter some fundamental concepts and definitions relevant to process development are provided, and our framework for designing a process investigation is shown.

## 2.1 A physicochemical approach to process development

Successful and efficient process development involves the study of a number of key aspects of the process, and allows design and development to be based on sound fundamental understanding rather than empirical rules of thumb. Key to achieving the objectives are an understanding of both the reaction and product recovery (workup) stages.

Much of the information required to understand the reaction stage can be gained by answering the following questions:

- **Reaction profiles** - how do the concentrations of reactants and products change during operation of the proposed process?
- **Pre-equilibria** - what equilibria affect the concentrations of the reacting species?
- **Competing reactions** - what are they, and what are their rates relative to that of the main reaction? What factors affect their rate?
- What **impurities** are present in the starting materials and what is their fate in the process?
- **Reactor configuration -** what is preferred?
- **Bulk and micromixing effects** - is the chemical selectivity dependent on reactor mixing and dynamics?
- **For multiphase systems** - What are the relevant distribution coefficients or solubilities? Is mass transfer or chemical reaction rate limiting? Is the agitation system providing adequate interfacial contact between the reacting phases?
- **Reaction modelling** - would a mathematical model help understanding?
- **Scale-up** - what additional understanding is necessary to ensure that the process will work on the manufacturing scale?

## 2.2 Reactor configuration

In the laboratory the most common method of conducting a reaction is to mix the reactants and wait for reaction to complete, often assisted by heating. This is termed batch, or sometimes 'all in' batch operation (Fig. 2.1). There are two main reasons why this method may not be feasible on the large scale:
1. For an exothermic reaction it may not be possible to remove the heat safely. This will depend on the heat and rate of reaction, and on the heat

removal capabilities of the vessel. Because the surface area : volume ratio is much less on the large scale (assuming same geometry), heat removal is likely to be much less efficient, unless special attention is paid to providing additional surface area.

2. If it is required to control a reaction parameter which changes as a result of chemical reaction e.g. pH in a reaction which generates acid, then a simple batch reaction is not the best way to proceed, because, in the event of any failure of the control or feed system, the reaction can no longer be controlled. In large scale chemistry it is essential that the chemist, not the chemicals, is in control.

When, as is common, batch operation is not feasible, the preferred method of operation is to add one (or more) reagents over a period of time to the substrate which is already in the vessel. This is known as *semi-batch* operation.

Choice of the order of addition is usually, but not always, fairly obvious. Where one reactant is unstable under the reaction conditions, but can be kept stable otherwise (a diazonium salt solution is one example), it should be kept under stable conditions and gradually fed to the reactor. If over reaction can occur by addition of one reagent to another, then the problem may be mitigated by reversing the order of addition.

An example is in the nitration of an organic substrate with nitric acid, using sulphuric acid to act as a desiccant in order to maintain the acidity of the reaction mass. Nitration reactions are widely used in the synthesis of aromatic amines. A typical example is shown in (Fig. 2.2).

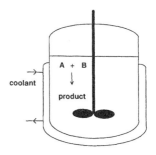

**Fig. 2.1** Batch or semi-batch reactor

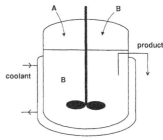

**Fig.2.3** Continuous stirred tank reactor

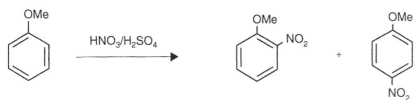

**Fig. 2.2** Nitration of anisole

They are highly exothermic reactions ($\Delta H \sim -170$ kJ mol$^{-1}$) and the adiabatic temperature rise is typically of the order of several hundred degrees. In order to achieve safe operation it is necessary to add the nitric acid slowly under conditions where it reacts immediately, so as to have a controlled heat output. Cooling is applied either to the reactor jacket, *via* an internal cooling coil, or both. With highly reactive substrates addition to the nitric acid is not possible: dinitration will usually occur, because the initially nitrated product is formed in the presence of a large excess of the nitrating species.

Batch or semi-batch reactors are very versatile, in the sense that they can be used for reactions with a wide range of chemical reaction rate constants, with the ability to carry out multiple consecutive operations in the same vessel. For this reason they are by far the most commonly used in multistage chemical manufacture. Continuous reactors may consist of a stirred vessel into which reactants are continuously fed and removed (Fig. 2.3), a series of such vessels, or a pipe into which reactants are fed at one end usually via a

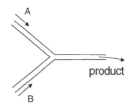

**Fig. 2.4** Tubular reactor

mixing device (Fig. 2.4). Where continuous reactors are used it is rare to have a multiproduct unit, because the residence time and mixing requirements usually vary too much to permit this, and because separate (although smaller) vessels are required for each separate process stage.

## 2.3   Yield, conversion and selectivity

Conversion means the fraction or percentage of the charged raw material which is consumed in the reaction. Chemical selectivity is the proportion of desired product formed from a given amount of starting material converted. The yield of a chemical reaction is the fraction or percentage of the desired product based on the amount of starting material charged. Definitions are shown in Table 2.2 for a reaction of **n** mols of starting material **A** which gives **p** mols of product **P** and **s** mols of by-product **S** (Table 2.1).

**Table 2.1**  Example process input and output

|        |       | A | $\rightarrow$ | P | + | S |
|--------|-------|---|---|---|---|---|
| Start  | moles | n |   | 0 |   | 0 |
| Finish | moles | a |   | p |   | s |

**Table 2.2**  Definitions of conversion, selectivity and yield

| Term | Definition |
|------|------------|
| Conversion | $100 \times (n-a)/n$ |
| Selectivity | $100 \times p/(n-a)$ |
| Yield | $100 \times (p/n)$ |

In the case where it is possible to recycle unreacted component A, the yield is equal to the selectivity. When recycle is not possible, and unreacted A is discarded, the yield is equal to $100 \times (p/n)$. For the selectivity to be less than 100% there must be some alternative reaction path (side reaction) to that desired.

## 2.4   Reaction profiles

Reaction profiling is a very powerful tool in process development. It is the measurement of the composition of a reaction for the reactive components, intermediates and products as the reaction proceeds. There are several forms that this information can take, depending on the reactor configuration. For a batch reaction, concentration *vs* time profiles are determined. For semi-batch reactions, which usually involve relatively rapid reactions, the reactions are substantially complete by the time addition of the reagent is complete. In these cases, concentration changes are measured versus the amount of the reactant which has been added. For continuous reactions, either plug flow or continuous stirred tank, the variable is residence time. For processes which are strongly catalysed, understanding the nature of the catalyst throughout the process may be a key part of this work. Not infrequently many subtleties are uncovered which are undreamed of at the start of the investigation.

## Some reaction profiles for batch reactions

Some idealised examples of reaction profiles are shown in Figures 2.5, 2.6 and 2.7 for 'all-in' (or batch) reactions. A simple case is a first order reaction proceeding cleanly to completion (Fig. 2.5). In this case, increasing the reactant concentration will have no effect on the time for a given proportion of the reactant to convert to product. This is rarely seen, but a second order reaction, where the rate is proportional to the concentration of each of two reactants is quite common. Second order reactions show a long 'tail' (Fig. 2.6), which in practice can be inconvenient. In this case, when equal concentrations of reactants are used, doubling the concentration will halve the time for a given fraction to react.

Use of quite a small excess of one component has a major effect on the time needed to complete a second order reaction (Table 2.3), and this technique is frequently used in order to achieve an acceptable reaction time, or, sometimes, to ensure that an expensive component is completely converted to product. The cheaper or more easily removable component will normally be used in excess.

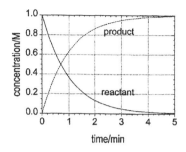

**Fig. 2.5** Concentration/time profile, first order reaction, rate = $k_1[A]$ $M^{-1}$ $min^{-1}$, $A_0 = 1$

**Table. 2.3** Effect of an excess of one reactant on time to react: second order reaction

| % A reacted | % excess of B | | |
|---|---|---|---|
| | 0 | 2 | 10 |
| | time to react (min) | | |
| 95 | 19 | 15.6 | 10 |
| 99 | 99 | 54 | 23 |

$k_2 = 1$ $M^{-1}$ $min^{-1}$; initial [A] = 1.0 M

Over-reaction of a desired product due to an excess of the reagent might normally be expected to be seen towards the end of a reaction, but this can be difficult to see from a concentration/time profile. Fig. 2.7 shows a typical example. Plotting the data in the form shown in Fig. 2.11 can be more informative.

When a product is formed *via* an intermediate then some knowledge of the possible side reactions of the intermediate is needed. For example, catalytic reduction of a nitrile is a method of producing primary amines. The process goes *via* an intermediate imine which can be diverted to a secondary amine by reaction with the desired primary amine (Fig. 2.8)

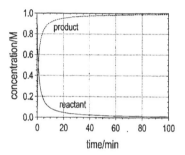

**Fig. 2.6** Concentration/time profile for second order reaction rate = $k_2.[A]^2$ $M^{-1}$ $min^{-1}$. $k_2 = 1$ $M^{-1}$ $min^{-1}$; initial. [A] = 1.0 M

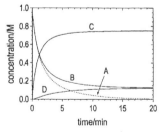

**Fig. 2.7** Concentration/time profile for a consecutive reaction. $k_1 = 0.9$ $M^{-1}$ $min^{-1}$; $k_2 = 0.1$ $M^{-1}$ $min^{-1}$ [A]$_0$ = [B]$_0$ = 1.0

$$A + B \xrightarrow{k_1} C$$
$$A + C \xrightarrow{k_2} D$$

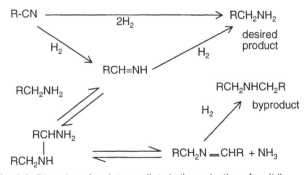

**Fig. 2.8** Diversion of an intermediate in the reduction of a nitrile

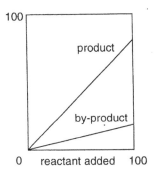

**Fig. 2.9** % Composition vs extent of reaction, by-product formed in parallel.

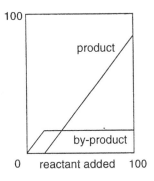

**Fig. 2.10** % Composition vs extent of reaction, impurity present in receiving reactant

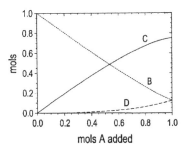

**Fig. 2.11** Composition *vs* extent of reaction, consecutive reaction (rates and stoichiometry as Fig. 2.7)

Special precautions have to be taken to avoid this. One method is to carry out the reduction in the presence of an excess of ammonia, so that the equilibrium preceding secondary amine formation is driven backwards.

**Reaction profiles for semi-batch reactions**

When a reaction is being operated under semi-batch conditions, reaction is occurring during the addition of reactants. The concept of a reaction profile is therefore different from that described for batch reactions. Nonetheless, useful information can still be gained from sampling semi-batch reaction. Two techniques can be used.

*Samples are taken at intervals throughout the addition and analysed.*

It is often found that the reaction selectivity changes throughout the addition of a reactant; this information can be just as useful as the reaction profiles shown for the batch reactions. These reaction profiles are plots of composition *vs* amount of reactant added.

By-product formed in parallel with the main product, such as an isomeric nitration product, will be found in constant ratio to the main product. Fig. 2.9 shows this case.

A common occurrence is poor selectivity early on in the reactant addition, caused by some impurity in the component in the vessel. The profile is shown in Fig. 2.10. For example, hydrolysis of a starting material due to a trace of water could be a cause of this problem. A consecutive reaction product formed by the further reaction of the desired product will normally increase only towards the end of the addition (component D in Fig 2.11), Unusual effects occur when the rate of *both* reactions are faster than the rate of mixing of the reactants. In this case by-product D will be seen throughout the reagent addition. This case could be confused with a parallel reaction but for the stoichiometry, which shows that something unusual is occurring. There is further discussion of this aspect in Chapter 5.

*Add reagent portionwise and examine the reaction profile after each addition*

This will immediately give a good idea of the reaction rate, which is useful for two reasons.

1. It shows whether there is any accumulation of reactant which could have hazard implications.

2. A very rapid reaction, occurring on the timescale of local or bulk mixing, may show a chemical selectivity different from that expected from those based on the kinetics of a perfectly mixed system (This is explained in Chapter 5).

**Catalytic reactions**

For a reaction which only proceeds in the presence of a catalyst, an understanding of how the catalyst works can be of great assistance in process optimisation. For example, in a Kharasch addition catalysed by a redox metal couple the catalysis is provided by the low valent metal (Fig. 2.12).

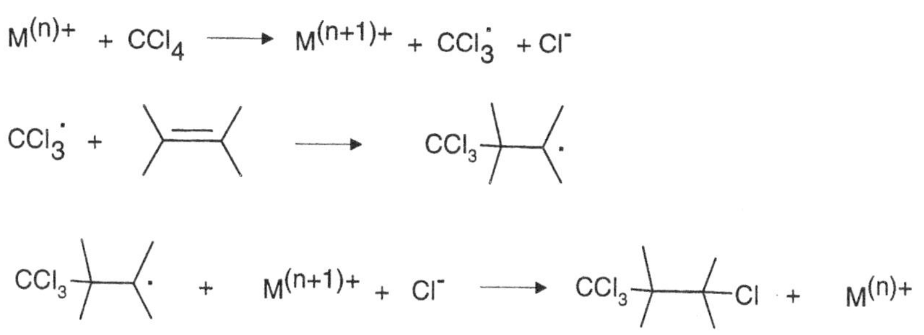

**Fig. 2.12** Catalytic cycle in a Kharasch reaction

The redox cycle is broken by side reactions such as radical dimerisation which means that the metal is gradually transformed from the catalytic low-valent state to the higher oxidation state, which is inactive as a catalyst. Once this is understood, methods can be found to ensure that a sufficient proportion of the metal remains in the low-valent state e.g. by adding a reducing agent.

### Sampling

In order to get useful information about a process it is necessary to take samples which represent the state of the reaction under investigation. This might seem obvious, but quench procedures need to be carefully thought out so as not to change the composition in a significant way. It is important not to put the samples through a workup procedure which could remove some components. It may not be possible to diagnose the state of a reaction mass reliably if, for example, the product is crystallised and isolated before analysis, as the isolation itself may remove impurities.

## 2.5 Purification and Work-up

There is a tendency to focus attention on the reaction stages in order to provide a high yield or highly selective process. The science involved in developing efficient workup processes is less familiar to many development chemists and engineers. Clearly a high selectivity is important. Minimising by-product formation keeps materials costs down and reduces waste which usually requires disposal to effluent treatment. In complex organic synthesis, a high conversion is also often important to achieve a high yield because of the difficulty of separating the reagents from the products, and recovering them for recycle. However, it should not be assumed that this is *always* the case. In large tonnage petrochemical and inorganic chemical manufacture high selectivity and low conversion are common with feed separation and recycle, particularly in continuous operation.

### To isolate or not?

When a chemical route to a complex organic molecule is invented by a research department, it will commonly have a large number of reaction stages, with purification of intermediates following each reaction stage. This is necessary because the reaction stages will not have been optimised. At each reaction stage there could be a large number of by-products produced by the large number of potentially reactive sites on the molecule, which complicates the separation. Many species present will be of unknown composition and may well have similar properties to the desired reaction product. The challenge for the development team is to produce a robust and operable *total process* to manufacture to the desired purity. The process includes all stages, both reaction and isolation, and should achieve the optimum cost. Note that the cost includes equipment (capital), materials and equally important, the cost of waste treatment. The process will also have to be registered under the Integrated Pollution Prevention and Control regulations and will therefore have strict limitations imposed upon allowable emissions.

It will therefore be important to *minimise* the number of purification stages as generally they produce waste streams for treatment or recovery, require a lot of expensive equipment and intermediate storage. They also almost invariably introduce a yield loss. It is not possible to give general guidance on when to use intermediate isolation. *The key to an informed decision is an understanding of the effect of impurities on downstream stages.* However it can be stated that, as opposed to route invention, the norm in development would be *not* to isolate unless absolutely necessary. The exception would be when deliberately using reaction conditions which give a low conversion, possibly to achieve a high selectivity, which would then require separation of reaction products and unreacted feeds for recycle.

**Selection of isolation technology**

A key part of process development is selection of the best isolation technology and the development of the isolation stages to be robust, cost effective and operable at full scale. The most common isolation technique used in research or route invention is crystallisation, often from a solvent, followed by filtration, washing and possibly further recrystallisation. Whilst it is a good technique for producing high purity, particularly with multiple recrystallisations and multiple solvents, it is one of the more expensive. It requires a lot of equipment, potentially incurs considerable yield loss and generates significant quantities of effluent or waste streams for treatment. Multiple crystallisation and washing also involves handling solids which is potentially problematic at large scale.

For robust operation, it is preferred to avoid handling solids, because solids handling equipment generally involves a lot of mechanical machinery which will require maintenance, large scale equipment trials and some uncertainty in how it will perform at full scale. The uncertainty comes from the interaction between the solid properties and the equipment performance. For example, the ability of a material to wash free of impurities will depend upon the physical form, which can be difficult to predict from small scale experiments. It could therefore be possible to have a very simple crystallisation in the laboratory followed by a filtration on a filter paper and a wash with solvent from a wash bottle, whereas at full scale the form produced could be very difficult to separate on realistic sized equipment, require extreme amounts of solvent to wash and involve a large yield loss.

**Table 2.4** Isolation technology options

| Technology | Comments |
|---|---|
| Distillation | Good provided the product and impurities have a significant difference in volatility and the distillation temperature and pressure are easily achievable. Also that the products are stable at the required temperatures for the distillation cycle time. Yield loss will depend upon stability and the relative volatility of the product and impurities. Will generate the waste as a concentrated liquid or residue for further recovery or disposal. |
| Liquid Extraction | Good provided the products and impurities have a significant difference in partition in readily useable solvents and that the solvents used can be recycled or recovered easily. Multiple stages can be used. Yield loss depends upon solubility. Will generate a waste effluent stream for treatment. Potentially can be made highly selective by extracting agents, ligands and the choice of appropriate ion-pairing agents. |
| Precipitation | Particularly useful for isolating organics from inorganics by precipitation from water, but requires filtration and washing. Generates effluent streams and can be very dependent on physical form. |
| Crystallisation | Can generate high purity and can be operated in a large number of ways. Potentially very versatile. Requires filtration and washing and can be very dependent on physical form. |

In order of preference from the point of view of equipment cost, robustness of operation, and potential for yield loss and generation of effluent, the preferred common isolation technologies are shown in Table 2.4.

There are several other isolation and purification technologies which can be useful in specific circumstances. Membranes can achieve separation by molecular size, for solutes, or particle size for suspensions and colloids. Adsorption and chromatography achieve selectivity in the same way as analytical chromatography, by selective molecular interaction with the adsorbent. Absorption can achieve selectivity from a gaseous or vapour phase by selective dissolution in the absorbing liquid.

**Final product purification**

The product will be required to meet a specification, which will be determined by the samples submitted for registration with the appropriate authorities. The specification will include reference to the chemical composition but may well also include physical aspects of the product, for example, in the case of a solid, polymorphic form and particle size. The registration could be based on samples generated for clinical trials, application trials or toxicological testing. It is important during development not to base the specification on samples which are unrealistically pure. Setting a specification which is more pure than the application requires may make it very difficult to manufacture in-specification product at full scale.

Earlier comments concerning technology selection apply to also to the final product purification. In this case however, it is likely that a combination of a number of technologies will be required to meet all aspects of the specification. When the product is a solid, a significant amount of attention will have to be paid to ensuring that the physical aspects of the specification are achievable. Pilot scale equipment trials will probably be required. These trials will require a significant amount of product and must be planned well in advance, especially for a multi-stage synthesis, in order that the material can be made available within the timescale of the development programme.

# 3 Pre-reaction equilibria

Equilibria involving either organic or inorganic species can play an important part in influencing both the reaction and workup stage of organic syntheses.

In the *reaction stage*, these equilibria can determine the proportion of *reactive* species present, which may not be the same as the total or stoichiometric concentration, and can thus influence both the rate and selectivity of the reaction. During *workup*, the same principles can determine the actual concentration of species involved in crystallisation or extraction processes.

## 3.1 Examples of the influence of pre-equilibria

### Reaction stage

When a reactive species is involved in a chemical equilibrium process which influences its concentration, the equilibrium process is referred to as a pre-reaction equilibrium or more commonly as a pre-equilibrium process. Common examples are shown in Table 3.1; some useful $pK_a$ values are given in Table 3.2.

**Table 3.2** Some useful $pK_a$ values

| Compound | $pK_a$ |
| --- | --- |
| $PhNH_3^+$ | 4.7 |
| $Et_3NH^+$ | 11.0 |
| Pyridinium ion | 5.25 |
| Phenol | 9.9 |
| Acetic acid | 4.75 |
| Nitrous acid | 3.4 |
| Hydrogen fluoride | 3.17 |

**Table 3.1** Some pre-reaction equilibria

| Equilibrium | Example |
| --- | --- |
| acid-base | $R\overset{+}{N}H_3 \rightleftharpoons RNH_2 + H^+$ |
|  | $ArOH \rightleftharpoons Ar\overset{-}{O} + H^+$ |
| nitronium ion formation | $HNO_3 + H^+ \rightleftharpoons NO_2^+ + H_2O$ |
| nitrosation catalysis | $HNO_2 + HX \rightleftharpoons NOX + H_2O$ |
| tribromide formation | $Br^- + Br_2 \rightleftharpoons Br_3^-$ |
| 'homoconjugation' | $HF + F^- \rightleftharpoons HF_2^-$ |

In all these cases the species on one side of the equilibrium have different reactivity to those on the other side, and consequently the overall reactivity of the system depends on the position of the equilibrium. An example is the coupling reaction between an aromatic diazonium salt and an aromatic amine (Fig. 3.1), which is industrially important as the route to the widely used azo dyes:

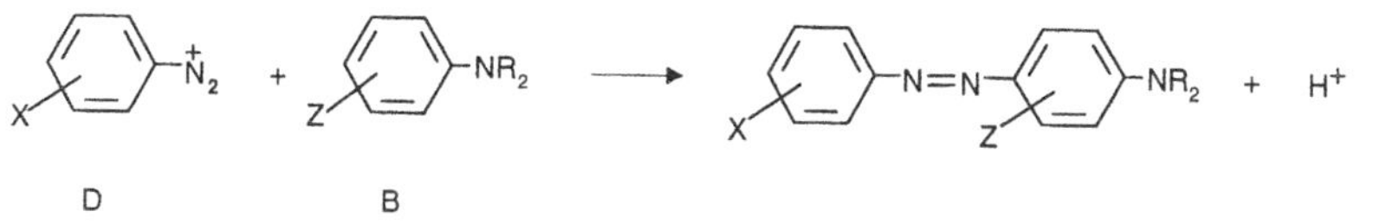

X is typically -NO$_2$, -Cl, or -H
Z is typically CH$_3$NHCO- or H-

**Fig.3.1** An 'azo-coupling' reaction.

These reactions are typically carried out at a pH in the region of the pK$_a$ of the conjugate acid of the aromatic amine, which is usually in the range 4 – 6 depending on the substituents. Reaction occurs between the free amine and the diazonium salt, so that the rate of the reaction is determined by the *actual* concentration of the free amine, which is determined by the position of the pre-equilibrium, and may not be the same as its *stoichiometric* concentration. Quantitative treatment is quite simple. The reaction rate is known to be separately proportional to the concentrations of    diazonium ion and unprotonated (free) amine (eqn. 3.1)

Over the pH range normally used for this coupling, the diazonium salt is almost entirely in the form of the diazonium ion shown.

$$\text{rate} = k_2[D][B] \tag{3.1}$$

where $k_2$ is the second order rate constant, [D] is the concentration of the diazonium ion and [B] the concentration of the free amine.

It is required to know the dependence of the reaction rate on the pH of the solution.

The relevant pre-equilibrium is (eqn. 3.2)

$$BH^+ \rightleftharpoons B + H^+ \tag{3.2}$$

and the equilibrium constant is defined in eqn. 3.3.

$$K_a = \frac{[B][H^+]}{[BH^+]} \tag{3.3}$$

A mass balance on the amine species gives eqn. 3.4

$$[B]_T = [BH^+] + [B] \tag{3.4}$$

where [B$_T$] is the stoichiometric concentration of amine.
By substituting eqn. 3.3 into this mass balance eqn. 3.5 is obtained

$$\frac{[B]}{[B]_T} = \frac{1}{\left(1 + \dfrac{[H^+]}{K_a}\right)} \quad \text{or} \quad \frac{[B]}{[B]_T} = \frac{K_a}{K_a + [H^+]} \tag{3.5}$$

Remember that pH = $-\log_{10}[H^+]$ and that pK$_a$ = $-\log_{10}K_a$

Note that when pH >> pK$_a$, eqn 3.5 reduces to [B] = [B]$_T$ and that when pH << pK$_a$ it reduces to [B] = [B]$_T$K$_a$/[H$^+$]

**Table 3.3** Fraction of total amine as free amine *vs* pH-pK$_a$

| pH-pK$_a$ | Fraction as free amine |
|---|---|
| –2.0 | 0.01 |
| –1.5 | 0.031 |
| –1.0 | 0.091 |
| –0.5 | 0.24 |
| 0 | 0.5 |
| 0.5 | 0.76 |
| 1.0 | 0.91 |
| 1.5 | 0.97 |
| 2.0 | 0.991 |

This equation shows the variation in the fraction of free base with [H$^+$]. Some values are shown in Table 3.3 and are plotted in Fig. 3.2 with [H$^+$] converted to pH.

Eqn. 3.5 occurs in all calculations where the change in a protonation equilibrium is involved. Fig. 3.2 shows the concentration/pH plot, which is useful in that it shows clearly the gross behaviour of the system.

14 *Pre-reaction equilibria*

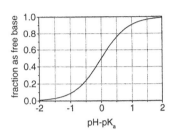

**Fig. 3.2** Plot of fraction as free base *vs* (pH-pK$_a$)

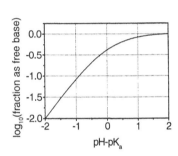

**Fig. 3.3** Plot of log$_{10}$(fraction as free base) *vs* (pH-pK$_a$)

pK$_a$ = 7.47

pK$_a$ = 9.1

Many reactions, however, occur *via* very low concentrations of reactive intermediates, and in this case it is helpful to use the logarithmic form of the plot as shown in Fig. 3.3. This shows more clearly that, when the pH is well below the pK$_a$, the concentration of free base increases by a factor of 10 for each one unit increase in the pH.

By substituting eqn 3.5 into eqn 3.1 the dependence of reaction rate on pH is readily obtained (eqn. 3.6), and is of the same form as Figs. 3.2 or 3.3.

$$\text{rate} = \frac{k_2[D][B]_T}{\left(1+\dfrac{[H^+]}{K_a}\right)} \tag{3.6}$$

The reaction rate reaches half its maximum value when the pH is equal to the pK$_a$ of the amine, and at one pH unit above the pK$_a$ has reached 91% of its maximum value. It is common to refer to the experimentally observed rate constant as k$_{obs}$ (or sometimes k$_e$). By inspection of 3.6 it can be seen that

$$k_{obs} = \frac{k_2}{\left(1+\dfrac{[H^+]}{K_a}\right)} \tag{3.7}$$

The same general principles apply to reactions in which more complicated but rapid pre-equilibria apply.

**Work-up processes**

Liquid/liquid extraction processes involving ionisable solutes and processes involving poorly soluble ionisable solids will show a sensitivity to the pH of the solution. These topics are treated in Chapter 6.

## 3.2 Some common simple pre-equilibria - influence on reaction rates

**Hypochlorite**

Cyanide ion is widely used as a reagent in the synthesis of organic chemicals, and because of its toxicity any excess must be destroyed, usually by oxidation, before discharge of effluent to drains or sewage works. Hypochlorite is the usual oxidant (eqns. 3.10 and 3.11). The acid form is much more reactive than the hypochlorite anion, so the reactivity towards cyanide ion is dependent on the pH, which influences the concentrations of the reacting species (eqns. 3.8, 3.9).

$$HOCl \rightleftharpoons ClO^- + H^+ \tag{3.8}$$

$$HCN \rightleftharpoons CN^- + H^+ \tag{3.9}$$

At high pH (>11) reaction of hypochlorous acid with cyanides occurs via the cyanide ion, not via hydrogen cyanide. Hypochlorous acid is much more reactive than the hypochlorite anion.

$$CN^- + ClO^- \xrightarrow{\ k_{ClO^-}\ } \text{products} \qquad (3.10)$$

$$k_{ClO^-} = 310 M^{-1} s^{-1}$$

(It is not clear whether the initial products are $Cl^-$ and $CNO^-$ (cyanate) or Cl-CN (cyanogen chloride)and $OH^-$, in which case a water molecule would also be involved).

$$CN^- + HOCl \xrightarrow{\ k_{HClO}\ } Cl - CN + OH^- \qquad (3.11)$$

$$k_{HClO} = 1.22 \times 10^9 M^{-1} s^{-1}$$

The overall rate expression for the destruction of cyanide ion is given by the eqn. 3.12.

$$\frac{d[CN^-]}{dt} = [CN^-]\{k_{ClO^-}[ClO^-] + k_{HClO}[HClO]\} \qquad (3.12)$$

This expression is not helpful as it stands, because the proportions of hypochlorite anion and hypochlorous acid vary with pH. In the alkaline pH range involved (pH 11 and above) it is only the variation in the concentration of hypochlorous acid which is important, since the concentration of the free hypochlorite and cyanide anions hardly changes, and free hydrogen cyanide is unreactive. By transforming the hypochlorous acid concentration into a term in hypochlorite using the expression for $K_{HClO}$ (eqn. 3.13) and the ionic product of water, $K_W = [H^+][OH^-]$, a more useful form is obtained, as follows:

$$K_{HClO} = \frac{[H^+][ClO^-]}{[HClO]} \qquad (3.13)$$

Equation (3.12) is thus transformed to 3.14.

$$\frac{d[CN^-]}{dt} = [CN^-][ClO^-]\left\{k_{ClO^-} + \frac{k_2}{[OH^-]}\right\} \qquad (3.14)$$

$$k_2 = 583 s^{-1}$$

so the observed rate constant is given by eqn. (3.15):

$$k_{obs} = \left\{k_{ClO^-} + \frac{k_2}{[OH^-]}\right\} \qquad (3.15)$$

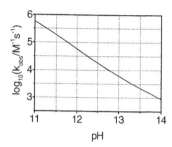

**Fig. 3.4** $k_{obs}$ *vs* pH in the region pH 11 to 14, calculated from (3.15)

$pK_W = 14$ in pure water at 25°C, but 13.79 at an ionic strength of 1.0.

Data from Gerritson C.M. and Margerum D.W (1990). *Inorg. Chem.*, 29, 2757.

Fig. 3.4 shows how the effective rate constant, $k_{obs}$, calculated from eqn.3.15, varies with pH. Below pH 11 an additional correction must be made for the removal of cyanide ion as HCN.

In considering appropriate detoxification conditions, hydrolysis of the cyanogen chloride must also be considered. This increases in rate with pH (Chapter 4, Table 4.4), and in the laboratory the recommended conditions for cyanide destruction are around pH 13–14, under which conditions cyanogen chloride is rapidly destroyed. Reaction of cyanide ion with hypochlorite at significantly lower pH can lead to escape of this toxic gas from the solution.

## Bromide/tribromide equilibrium

The rate of bromination of stilbene in methanol is *decreased* by the addition of bromide. This is because free bromine is removed as tribromide (eqn. 3.16), which is less reactive than free bromine. The formation constant is given by eqn.3.17.

Stilbene is:

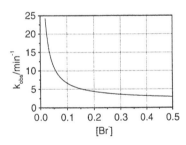

$$Br^- + Br_2 \rightleftharpoons Br_3^- \tag{3.16}$$

Note that this equilibrium constant refers to methanol as solvent. Data from Bartlett P.D. and Tarbell D.S. (1936). *J. Amer. Chem. Soc.*, 58, 466.

$$K_{Br_3^-} = \frac{[Br_3^-]}{[Br^-][Br_2]} = 417 M^{-1} \tag{3.17}$$

The rate of bromination of stilbene under the conditions employed can be fitted to a model which includes bromination by both tribromide and free bromine: (eqn.3.18)

$$k_{Br_2} = 200 M^{-1}s^{-1}$$
$$k_{Br_3^-} = 2.02 M^{-1}s^{-1}$$

$$rate = k_{Br_2}[Br_2][S] + k_{Br_3^-}[Br_3^-][S] \tag{3.18}$$

where [S] is the concentration of stilbene. Kinetic measurements were carried out under conditions where bromide ion was in large excess over bromine. Under these conditions the bromide concentration can be regarded as constant and the relevant mass balance is given by eqn. 3.19, where $[Br_2]_T$ is the total amount of bromine present.

Subscript T is used to indicate the total i.e. stoichiometric concentration of the compound in question.

$$[Br_2]_T = [Br_2] + [Br_3^-] \tag{3.19}$$

Using (3.17) and (3.19) the rate expression (3.18) transforms to eqn. 3.20

$$rate = k_{obs}[Br_2]_T[S] \tag{3.20}$$

where $k_{obs}$ is given by eqn.3.21.

$$k_{obs} = \frac{k_{Br_2} + k_{Br_3^-} K_{Br_3^-}[Br^-]}{1 + K_{Br_3^-}[Br^-]} \tag{3.21}$$

Fig. 3.5 shows the variation in the observed rate constant $k_{obs}$ with bromide concentration calculated from eqn. 3.21. Note that the rate constant drops very rapidly as the bromide concentration is increased, and the reaction soon becomes dominated by bromination via the tribromide anion This may be useful if it is desired to moderate the reactivity of the system to avoid selectivity problems caused by rapid reactions (see Chapter 5).

**Fig. 3.5** Effect of bromide concentration on the rate of bromination of stilbene in methanol at 0°C.

## 3.3   Multiple ionisation in aqueous systems

### Linear coupled equilibria

When developing processes involving di-, tri- and poly-acidic species it is important to understand their complex but calculable ionisation behaviour.

$$H_2A \underset{}{\overset{K_1}{\rightleftharpoons}} HA^- \underset{}{\overset{K_2}{\rightleftharpoons}} A^{2-} \qquad (3.22)$$

For a diacid (which could be a diphenol or an diammonium salt) calculation of the ionisation curves follows the same principles as before: the mass balances and equilibria are written down (eqns.3.23–3.25):

$$[HA]_T = [H_2A] + [HA^-] + [A^{2-}] \qquad (3.23)$$

$$K_1 = \frac{[HA^-][H^+]}{[H_2A]} \qquad (3.24)$$

$$K_2 = \frac{[A^{2-}][H^+]}{[HA^-]} \qquad (3.25)$$

Solving these for the fraction of the three species in solution in terms of $[H^+]$ is done by successively substituting (3.24) and (3.25) into (3.23) to give eqns. 3.26–3.28.

$$\frac{[H_2A]}{[HA]_T} = \frac{1}{1 + \dfrac{K_1}{[H^+]} + \dfrac{K_1 K_2}{[H^+]^2}} \qquad (3.26)$$

$$\frac{[HA^-]}{[HA]_T} = \frac{\dfrac{K_1}{[H^+]}}{1 + \dfrac{K_1}{[H^+]} + \dfrac{K_1 K_2}{[H^+]^2}} \qquad (3.27)$$

$$\frac{[A^{2-}]}{[HA]_T} = \frac{\dfrac{K_1 K_2}{[H^+]^2}}{1 + \dfrac{K_1}{[H^+]} + \dfrac{K_1 K_2}{[H^+]^2}} \qquad (3.28)$$

The second and third expressions can be simplified by dividing through by the numerator, but the reason for showing them in the above form is that the symmetry can be seen, which enables the expression to be written down directly for analogous case where there are four or five species involved in the same linear relationship. Fig. 3.6 shows the distribution of the three species vs pH for malonic acid, where the difference in $pK_a$ between the two acid groups is 4.3. Whether or not the monoanion can be obtained as the only

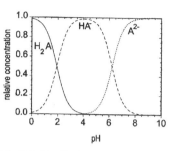

Fig.3.6 Distribution of species *vs* pH for malonic acid.  $pK_1 = 1.92$.  $pK_2 = 6.23$

Malonic acid is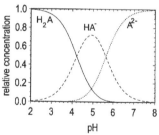

In this example, if $HA^-$ were the reactive species then it would be important to operate at pH 4. This Figure could be used to obtain the rate penalty for moving away from the optimum pH.

These terms each show the amount of the species shown in the numerator as a fraction of the total.

Fig. 3.7 Distribution of species *vs* pH for succinic acid..  $pK_1 = 4.21$. $pK_2 = 5.64$

Succinic acid is

species in solution at a particular pH depends on the difference in $pK_a$ between the neighbouring species. Fig. 3.7 shows the calculated species distribution for succinic acid, where the difference is only 1.4 $pK_a$ units.

These equilibrium calculations can quickly be solved using a programmable calculator, a simple spreadsheet calculation, or by any of several modelling packages which are commercially available.

## 3.4   Homoconjugation

The situation where a product of an equilibrium reacts with the starting material is referred to as homoconjugation. The best known examples in water are the hydrogen fluoride/bifluoride (eqns 3.29 and 3.30) and acetate/carboxylic acid equilibria (3.31). These complexes are quite weak, in contrast to those formed in non-aqueous solvents, where solvation is much stronger.

$$HF \xrightleftharpoons{K_{HF}} H^+ + F^- \tag{3.29}$$

$$HF + F^- \xrightleftharpoons{K_{HF_2^-}} HF_2^- \tag{3.30}$$

For carboxylic acids:

$$RCO_2H + RCO_2^- \xrightleftharpoons{K_{dimer}} (RCO_2^- .... HO_2C.R) \tag{3.31}$$

Organofluorine compounds are important in many product application areas. Their processing can generate small amounts of hydrogen fluoride. Both hydrogen fluoride and the bifluoride ion are corrosive to glass, which is commonly used to line process vessels. Fluoride ion itself is very much less corrosive. Formation of bifluoride ion means that corrosion occurs at a higher pH than would be the case if hydrogen fluoride alone was the corrosive agent (Fig. 3.8). To treat this case quantitatively the procedure is, as always, to write down the mass balance and equilibria (eqns. 3.32–3.34):

Care is needed with the mass balance. Note that bifluoride contains two mols of F

$$[F]_T = [HF] + 2[HF_2^-] \tag{3.32}$$

$$K_{HF} = \frac{[H^+][F^-]}{[HF]} \tag{3.33}$$

$K_{HF_2^-} = 8M^{-1}$

$$K_{HF_2^-} = \frac{[HF_2^-]}{[HF][F^-]} \tag{3.34}$$

Solving these equations gives a quadratic (eqn. 3.35) which can be solved in the normal way.

$$[HF]^2 + [HF]\left\{\frac{[H^+]}{2K_{HF_2^-}K_a} + \frac{1}{2K_{HF_2^-}}\right\} - \frac{[HF]_T[H^+]}{2K_{HF_2^-}K_a} = 0 \qquad (3.35)$$

Then $[F^-]$ and $[HF_2^-]$ are calculated from the equilibrium expressions. Use of a spreadsheet takes much of the tedium out of these calculations. Note that the species distribution now depends on the total fluoride concentration as well as the solution pH. Fig. 3.8 shows two examples. The fluoride concentrations are omitted for clarity.

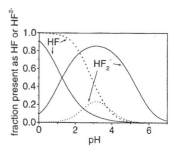

Fig. 3.8 Composition of aqueous HF/F⁻ solutions *vs* pH
$\cdots\cdots$ $[HF]_T = 0.1M$
——— $[HF]_T = 1M$

## 3.5 Other coupled equilibria

Often both starting materials and products are involved in acid/base equilibrium processes. Cyanohydrins are useful intermediates in the manufacture of agrochemicals and pharmaceuticals, and are formed from aldehydes and hydrogen cyanide (Fig. 3.9).

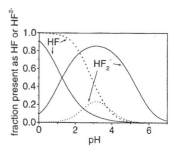

(BA)　　　　　　　　(CA)　　　　　　　　(CH)

Fig. 3.9 Cyanohydrin equilibria.

The rate and equilibrium process involved in cyanohydrin formation are very sensitive to pH and to the concentrations of reactants, and it is important to understand these in order to optimise the process conditions for their formation and use. Three equilibria are involved: formation of the cyanohydrin, ionisation of hydrogen cyanide and ionisation of the cyanohydrin. We wish to calculate the fraction of aldehyde converted to cyanohydrin *vs* pH and the concentration of cyanide used. Quantitative treatment follows the usual system. Equilibria (eqns 3.36–3.38) and mass balances (eqns. 3.39–3.41) are written down and solved with pH as the independent variable.

$$BA + HCN \underset{}{\overset{K_F}{\rightleftharpoons}} CH \qquad (3.36) \qquad K_F = 236M^{-1}$$

$$HCN \underset{}{\overset{K_{HCN}}{\rightleftharpoons}} H^+ + CN^- \qquad (3.37) \qquad pK_a = 9.1$$

$$CH \underset{}{\overset{K_{CH}}{\rightleftharpoons}} H^+ + CA \qquad (3.38) \qquad pK_a \text{ for benzaldehyde cyanohydrin} = 10.73$$

where BA = benzaldehyde, CH = cyanohydrin and CA = cyanohydrin anion.

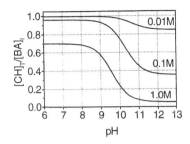

**Fig. 3.10** Plot of $\frac{[CH]_T}{[BA]_I}$ vs pH. for benzaldehyde cyanohydrin formation, for the three different values of $[HCN]_T$ shown.

Data from Kallen R.G. and Ching W.M. (1978). *J. Amer. Chem. Soc.*, 100, 6119.

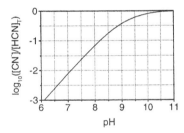

**Fig. 3.11** Plot of $\log_{10}([CN^-]/[HCN]_T)$ for HCN

Calculation is greatly simplified by assuming a fixed concentration of total cyanide in solution after equilibrium has been achieved *i.e.* cyanide in excess over the aldehyde. Mass balances are then

$$[BA]_I = [BA] + [CH] + [CA] \tag{3.39}$$

$$[CH]_T = [CH] + [CA] \tag{3.40}$$

$$[HCN]_T = [HCN] + [CN^-] \tag{3.41}$$

Eqn. 3.40 is divided by (3.39) and solved for the stoichiometric terms using the equilibrium constants (3.36, 3.37 and 3.38), to give the proportion of total aldehyde present as cyanohydrin (eqn. 3.42).

$$\frac{[CH]_T}{[BA]_I} = \frac{\left\{1 + \dfrac{K_{CH}}{[H^+]}\right\}}{\left\{1 + \dfrac{K_{CH}}{[H^+]}\right\} + \dfrac{1 + \dfrac{K_{HCN}}{[H^+]}}{K_F[HCN]_T}} \tag{3.42}$$

This function is plotted in Fig. 3.10 for benzaldehyde. It is noteworthy that there are two 'equilibrium constants', one at low pH corresponding to the addition of free HCN to the aldehyde, and another at high pH corresponding to the addition of cyanide ion to give the cyanohydrin anion, which is much smaller. Although the *equilibrium constant* is most favourable at a pH less than 7, the *reaction rate* is proportional to the free cyanide concentration, and so the *rate* of formation of the cyanohydrin increases with pH. Fig. 3.11 shows a plot of $\log_{10}([CN^-]/[HCN]_T)$ *vs* pH. Well below the $pK_a$ of HCN the reaction rate thus increases by a factor of 10 for each 1 unit increase in pH. When considering the design of a process to form a cyanohydrin this is a major consideration: on going from pH 7 to pH 1 the reaction rate decreases by a factor of $10^6$.

### 3.6   More complex systems - Nitrosation and diazotisation

Nitrosation processes are strongly catalysed by a number of species - halides, thiocyanates and thiourea - which are themselves nitrosated rapidly in a favourable pre-equilibrium and then rapidly nitrosate the substrate. From a process development viewpoint it is necessary to understand these pre-equilibria because they strongly affect the rate of the overall process. Overall the nitrosation reaction is exemplified by Fig. 3.12:

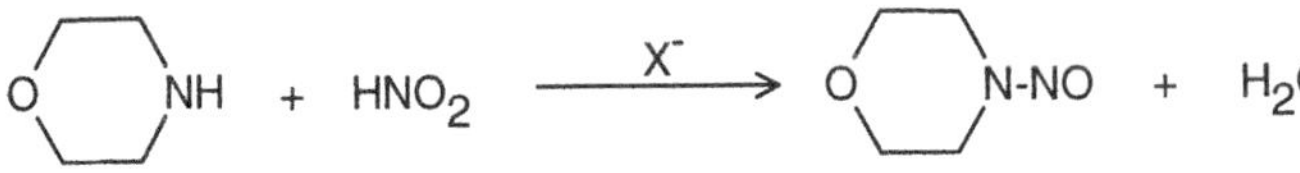

**Fig. 3.12** Nitrosation of morpholine

and the diazotisation process by Fig. 3.13.

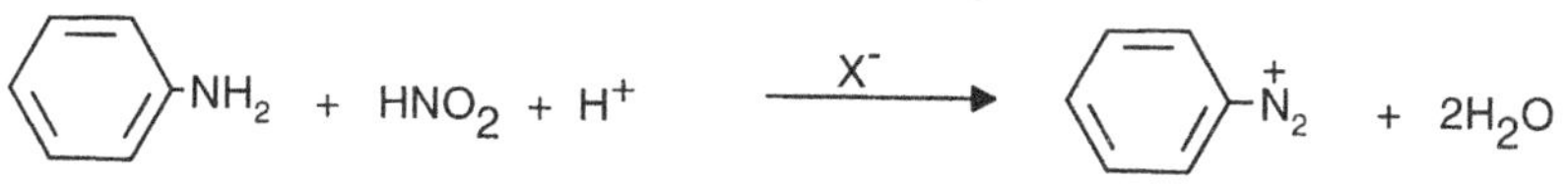

**Fig. 3.13** Diazotisation of aniline

Under conditions typically used the rate of diazotisation of aniline is given by eqn. (3.43)

$$rate = k_d[PhNH_2][X-NO] \tag{3.43}$$

For amines whose conjugate acids have a pKa < 4, the nitrosation step may be reversible and the rate expression more complex.

This cannot be used directly because $[PhNH_2]$ and $[X\text{-}NO]$ are not immediately known from the amounts of added of added reactants (the stoichiometric concentrations). The concentration of free aniline is easily calculated. Diazotisation is carried out at a pH in the region 1–2, which is well below the pKa of the anilinium ion. Therefore the simplified form of eqn. 3.5 can be used (see explanatory note to eqn. 3.5), to obtain (3.44)

$$[PhNH_2] = [PhNH_2]_T K_a / [H^+] \tag{3.44}$$

Anilinium ion pKa = 4.7

The pre-equilibrium for the nitrosating agent is given by eqn. 3.45.

$$HNO_2 + X^- + H^+ \underset{\longleftarrow}{\overset{K_{X\text{-}NO}}{\rightleftharpoons}} X-NO + H_2O \tag{3.45}$$

for which the equilibrium constant is

$$K_{X-NO} = \frac{[X-NO]}{[HNO_2][X^-][H^+]} \tag{3.46}$$

Since the pH is well below the pKa of nitrous acid, its stoichiometric concentration can be used to calculate the amount of X-NO derived from it. It is then assumed that the catalyst is in excess over the nitrous acid, so that [X⁻] is constant (compare the bromination example). The mass balance in nitrous acid is given by eqn. 3.47.

Nitrous acid pKa = 3.4

$$[HNO_2]_T = [HNO_2] + [X-NO] \tag{3.47}$$

Diazotisations are normally carried out by adding nitrous acid slowly to a solution of the amine in dilute acid containing the catalyst. When HCl is used as the acid, this also provides the chloride catalyst.

Combining eqns. 3.46 and 3.47 gives eqn. 3.48.

$$[X-NO] = \frac{K_{X-NO}[HNO_2]_T[H^+][X^-]_T}{1 + K_{X-NO}[H^+][X^-]_T} \tag{3.48}$$

Now the rate expression in terms of stoichiometric concentrations of reactants can be obtained by substituting eqns. 3.44 and 3.48 into eq. 3.43 to give eqn. 3.49:

$$rate = \frac{k_d K_{X-NO} K_a [HNO_2]_T [X^-][PhNH_2]_T}{1 + K_{X-NO}[H^+][X^-]_T} \tag{3.49}$$

These relationships come to life if they are used to explore the effect of the catalysts on the diazotisation rate. Key to the form of eqn 3.49 is the value of the term $K_{X-NO}[H^+][X^-]$ in the denominator. If this is much less than 1, as is the case for chloride, bromide and thiocyanate under the diazotisation conditions commonly employed, then eqn 3.49 reduces to the numerator, and the reaction rate is pH independent. In the case of thiourea this is not true and the rate is pH dependent (Fig. 3.14).

The $k_{calc}$ values (Table 3.4) give an indication of the relative catalytic efficiencies of the species shown under the particular set of conditions chosen. This system is particularly complex, with several pre-equilibria influencing the overall reaction rate. It is difficult to compare directly these rates with that of the uncatalysed reaction, which goes via dinitrogen trioxide ($N_2O_3$), because in that case the rate is limited both by the low equilibrium constant for formation of $N_2O_3$, and by its rate of formation (Bunton C.A., Llywellyn D.R. and Stedman G. (1959). *J. Chem. Soc.*, 568).

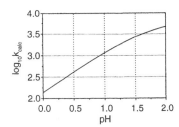

**Fig. 3.14** Plot of $k_{calc}$ *vs* pH for catalysis of the diazotisation of aniline by thiourea

$k_{calc}$ defined by:
rate $= k_{calc}[HNO_2]_T[PhNH_2]_T$

**Table 3.4** The equilibrium constant for formation of the catalysts X-NO, the second order rate constants $k_d$ for the diazotisation of aniline according to eqn. 3.43, and the rate constant $k_{calc}$, for diazotisation of aniline at pH 1, with the catalyst concentration set at $10^{-2}$ M

| Catalyst | $K_{X-NO}$  $M^{-1}$ | $k_d/M^{-1}s^{-1}$ | $k_{calc}/M^{-1}s^{-1}$ | $T/^{\circ}C$ |
|---|---|---|---|---|
| Cl$^-$ | $5.6 \times 10^{-4}$ | $2.5 \times 10^9$ | 0.28 | 25 |
| Br$^-$ | $5.1 \times 10^{-2}$ | $1.7 \times 10^9$ | 17.3 | 25 |
| SCN$^-$ | 30 | $2.9 \times 10^7$ | $1.7 \times 10^2$ | 31 |
| $(H_2N)_2C=O$ | 5000 | $7 \times 10^6$ | $1.17 \times 10^3$ | 31 |

Calculated from data in Crampton M.R., Thompson J.T. and Williams D. Lyn H. (1979). J. Chem. Soc., Perkin II, 18, and Meyer T. A. and Williams D. Lyn H. (1981), J. Chem. Soc., Perkin II, 361.

## 3.7   Summary

This chapter has exemplified a wide range of pre-reaction equilibria. Although the subject may appear a little 'dry' at this point, the significance of these equilibria should become more apparent later. Familiarity with these concepts is essential to the understanding of Chapters 4 and 6.

## Bibliography and sources of $pK_a$ data

Williams D.L.H.(1988). *Nitrosation*, Cambridge University Press.
Sergeant E.P. and Dempsey B. (1979). Ionisation Constants of Organic Acids in *Aqueous Solution, IUPAC Chemical Data Series - No 23*, Pergamon.
Perrin D.D., Dempsey B. and Sergeant E.P. (1981). *pK_a prediction for organic acids and bases*, Chapman and Hall.
Martell A.E.(1964). Stability Constants of Metal–Ion Complexes, *Chemical Society Special Publication No. 17*.
Christensen J.J., Hansen L.D. and Izatt R.M. (1976). *Handbook of Proton Ionisation Heats*, Wiley-Interscience.

# 4 Competing reactions in homogeneous systems

In process development a good yield is often key to obtaining satisfactory material costs, purity and robust operation. The essence of obtaining a good chemical yield is the understanding and control of those processes which compete with the desired reaction. This understanding is necessary to obtain a satisfactory laboratory process and, even more so, to scale up the process for manufacture (see Chapter 10).

## 4.1 Types of competing processes

Some types of competing reaction are exemplified in Table 4.1 and a generalised scheme is shown in Fig 4.1

**Table 4.1** Some types of competing reactions

| Main reaction | Competing | Examples |
|---|---|---|
| A → P | A → D | Decomposition of a diazonium salt to undesired by-product in parallel with product formation (Fig. 4.2). |
| A + B → P | P → D | Rearrangement of a kinetically controlled product to a thermodynamic product (Fig. 4.3). |
| A + B → P | A + solvent → D | Acylation of an amine with parallel electrophile solvolysis (Fig 4.4). Azo coupling with parallel solvolysis of the diazonium salt. |
| A + B → P | A + B → D | Aromatic substitution with isomer formation (Fig. 2.2). |
| A + B → P | B + P → D | Nitration followed by unwanted dinitration (Fig. 5.2). Monoacylation of a diamine with undesired diacylation (Fig 5.1). |
| A + B → Int Int + B → P | A + Int → D | Hydrogenation of a nitrile with competing secondary amine formation (Fig. 2.8). |

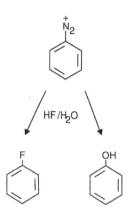

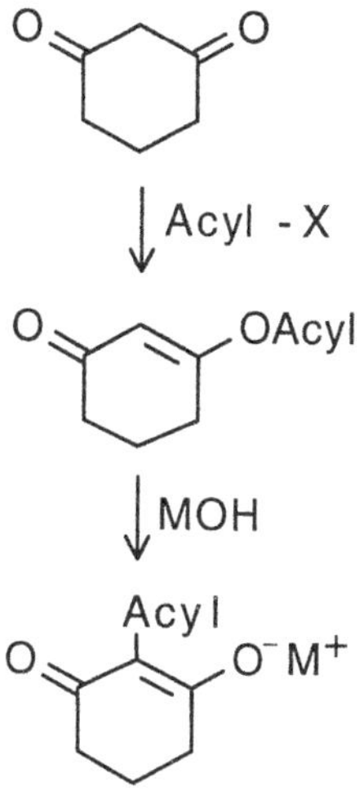

**Fig. 4.2** Decomposition of a diazonium salt in aqueous hydrogen fluoride. Nitrogen gas is liberated.

**Fig.4.3** Kinetic *vs* thermodynamic control in acylation of a 1,3-diketone

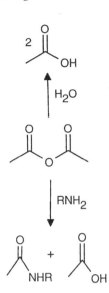

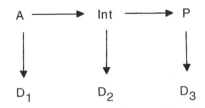

Int is an intermediate
$D_{1-3}$ are by-products

**Fig. 4.1** Possible side reactions in a transformation

Table 4.1 is a very simplified version of many real situations, because it does not display the many pre-equilibrium possibilities which were discussed in the previous chapter.

**Fig. 4.4** Acylation of an amine with competing solvolysis of the anhydride

## 4.2 Strategies for minimising side reactions

A pre-requisite for minimisation of side reactions is to know what they are. The basic investigative methods were outlined in Chapter 2. It is not sufficient to consider only the case of competitive homogeneous batch reactions, as may be implied by a reading of Table 4.1. Many selectivity problems occur in situations were the reactants are not fully mixed. This is the subject of Chapter 5. Some examples of both types are listed in Table 4.2

**Table 42**   Some problem situations and possible remedies

| Situation | Possible remedies |
|---|---|
| Reactant A unstable under reaction conditions | Add A to other reactant(s) in order to minimise the standing concentration of A and thereby to maximise the selectivity. Manipulate pre-equilibria. |
| Both reactants unstable under reaction conditions | Separate simultaneous feeds of both reactants to reactor. |
| Product intrinsically unstable under reaction conditions | Probably requires a low residence time plug-flow continuous process, or continuous removal of product., e.g. by distillation, extraction into another phase, or precipitation. |
| Product further reacts with a reagent | Depends on the timescale of the secondary reaction. May require the use of one reactant in excess and recycle. Manipulate pre- and post-reaction equilibria. |
| Loss of catalytic activity | In order to provide a solution it is necessary to understand the mode of action of the catalyst. |
| Selectivity requires a minimum concentration of one reagent. (Common in hydrogenations) | Ensure adequate reagent supply - this may be difficult in multi-phase systems.  Divert the reactive intermediate or product. |

The commonest parameters used to manipulate absolute and relative reactivities are concentration, order of reactant addition and temperature. Probably the major practical significance of temperature in influencing selectivity is the ability to avoid decomposition of an unstable material - reactant, intermediate or product - so as to permit its use on a sensible timescale. An example is the processing of a reactive intermediate where, in batch or semi-batch operation, it is necessary to maintain a low temperature in order to prevent decomposition on the processing timescale. It may be possible, by operating in continuous mode with a short residence time, to make and consume the intermediate at a much higher temperature. On the large scale this can lead to major cost savings. The use of temperature to alter relative reactivities in fully mixed homogeneous systems is less common.

## 4.3 Parallel/consecutive reactions

An important case is the selective reaction at one site in a bifunctional molecule, where, as the more reactive group is consumed, the ratio of the two groups changes, so that the selectivity falls off as the reaction proceeds. In the following example it is desired to monoacylate a diamine. Fig. 4.5 shows the general case wherein the amino groups **-NH$_2$** and **-NH$_2$** are of different reactivity.

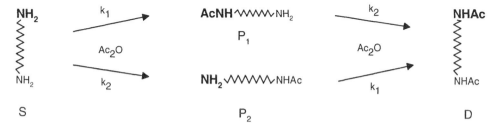

**Fig. 4.5** Acylation of a diamine

To illustrate the development problem consider the example where the relative rates of acylation to give $P_1$ and $P_2$ are in the ratio 1:9, and that the reactivity of the unreacted amine group in $P_1$ or $P_2$ is unaffected by acylation of the other amino group in the molecule. The rate expressions are given by eqns. 4.1–4.5:

$$\frac{d[Ac_2O]}{dt} = -(k_1 + k_2)[Ac_2O][S] - k_1[Ac_2O][P_2] - k_2[Ac_2O][P_1] \quad (4.1)$$

$$\frac{dS}{dt} = -(k_1 + k_2)[S][Ac_2O] \quad (4.2)$$

$$\frac{dP_1}{dt} = k_1[S][Ac_2O] - k_2[P_1][Ac_2O] \quad (4.3)$$

$$\frac{dP_2}{dt} = k_2[S][Ac_2O] - k_1[P_2][Ac_2O] \quad (4.4)$$

$$\frac{dD}{dt} = k_2[P_1][Ac_2O] + k_1[P_2][Ac_2O] \quad (4.5)$$

Table 4.3 and Fig. 4.6 show the product distribution at various ratios $R$ of Ac$_2$O to diamine.

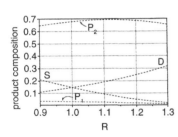

**Fig. 4.6** Product distribution (mol fraction) *vs* mols of acylating agent.

**Table 4.3** Product composition at various ratios *R* of Ac$_2$O to diamine

| R | S | P$_1$ | P$_2$ | D |
|---|---|---|---|---|
| 0.1 | 0.9 | 0.01 | 0.089 | 0.001 |
| 0.9 | 0.21 | 0.035 | 0.646 | 0.109 |
| 1.0 | 0.145 | 0.03 | 0.68 | 0.145 |
| 1.1 | 0.09 | 0.025 | 0.696 | 0.189 |
| 1.2 | 0.047 | 0.017 | 0.689 | 0.247 |
| 1.3 | 0.019 | 0.009 | 0.654 | 0.318 |

These calculations can be used to explore the processing options and to decide on the most appropriate reactant molar ratio. A number of factors may influence this choice, including the ease of recovery of unreacted material, the tolerance of the end use to the various by-products, and the relative material costs. It is evident that the selectivity is best with a large excess of the amine, but, unless it is readily separable, it is unlikely that this option will be economically viable.

### 4.4    pH-rate and selectivity profiles

King J. F., Rathore R., Lam J.Y.L., Guo Z.R. and Klassen D.F. (1992). J. Amer. Chem. Soc., 114 (8), 3028.

A common situation in aqueous reaction systems is competing reactions where solvolysis of a reagent, usually an electrophile **E**, competes with the desired reaction with a nucleophilic species **Nuc**. The reaction rate depends on the pH *via* the effect that this has on the availability of the reactive form of the nucleophile, and the selectivity depends on pH *via* the effect that this has on the electophile hydrolysis. Examples are the alkylation or acylation of phenols or amines (e.g. Fig. 4.4), or the C-condensation of aromatic amines with diazonium salts (Fig 3.1), where parallel decomposition of the diazonium salt also occurs. It is necessary to choose the appropriate pH range in order to optimise the process performance with respect to the desired reaction

Solvolysis of the electrophile usually proceeds *via* both direct reaction with water and by a stoichiometric reaction with hydroxide ion (eqn. 4.7)

$$r_{solv} = k_{H_2O}[E] + k_{OH^-}[E][OH^-] \tag{4.7}$$

so that the rate constant for the solvolysis $k_{solv}$ is given by eqn. 4.8.

$$k_{solv} = k_{H_2O} + k_{OH^-}[OH^-] \tag{4.8}$$

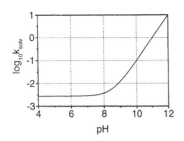

**Fig. 4.7** $\log_{10} k_{solv}$ *vs* pH for acetic anhydride.

As the pH increases the rate remains essentially constant until the point where the term in hydroxide ion becomes of similar magnitude to the water term. It then increases and eventually becomes dominated by the [OH$^-$] term. Here the pH at which $k_{H_2O} = k_{OH^-}[OH^-]$ is referred to as pK$_i$. Fig. 4.7 shows the plot of $k_{solv}$ *vs* pH for acetic anhydride, and Table 4.4 gives values of $k_{H_2O}$, $k_{OH^-}$ and pK$_i$ for some commonly used electrophiles.

**Table 4.4** $k_{H_2O}$ and $k_{OH^-}$ for some useful electrophiles (25°C)

| Electrophile | $k_{H_2O}$ /s$^{-1}$ | $t_{1/2}$ /s (H$_2$O) | $k_{OH^-}$ /M$^{-1}$s$^{-1}$ | pK$_i$ | ref. |
|---|---|---|---|---|---|
| COCl$_2$ | 6.0 | 0.12 | $1.6 \times 10^4$ | 10.6 | a |
| PhCOCl | 1.4 | 0.5 | 400 | 11.5 | b |
| PhSO$_2$Cl | $3.06 \times 10^{-3}$ | 226 | 40.4 | 9.88 | b |
| Ac$_2$O | $2.8 \times 10^{-3}$ | 248 | 970 | 8.5 | b |
| EtOCOCl | $4.85 \times 10^{-4}$ | 1430 | 40.6 | 9.08 | b |
| Cl–CN | $2.58 \times 10^{-6}$ | $2.69 \times 10^5$ | 4.53 | 7.75 | c |

[a] Manogue W.H. and. Pigford (1960). A.I.Ch.E. Journal, 6 (3), 494.
[b] King J. F. (1992). J. Amer. Chem. Soc., 114 (8), 3028.
[c] Bailey P.L. and Bishop E. J. (1973). J. Chem. Soc., Dalton Transactions, 912. (data at 26.5°C)

Product formation typically occurs *via* a second order process involving the electrophile and the basic form of the nucleophile, **Nuc**. The pre-equilibrium involving the base may be represented by eqn. 4.9 (cf Chapter 3).

$$NucH^+ \rightleftharpoons Nuc + H^+ \qquad (4.9)$$

Using the same terminology as Chap 3, the fraction of the nucleophilic species present in the unprotonated form is given by eqn. 4.10,

$$[Nuc] = \frac{[Nuc]_T}{\left(1 + \frac{[H^+]}{K_N}\right)} \qquad (4.10)$$

where $K_N$ is the dissociation constant for the protonated nucleophile. This is plotted in Fig. 4.8 for aniline. Thus the rate of electrophile hydrolysis and the availability of reactive free nucleophile have different dependencies on pH.

The rate of product formation is given by 4.11,

$$r_p = k_p[E][Nuc] = \frac{k_p[E][Nuc]_T}{\left(1 + \frac{[H^+]}{K_N}\right)} \qquad (4.11)$$

so for fixed [Nuc] i.e. reactant **Nuc** in large excess over E, the effective first order rate constant for product formation changes with pH as shown by eqn. 4.12

$$k_e = \frac{k_p[Nuc]_T}{\left(1 + \frac{[H^+]}{K_N}\right)} \qquad (4.12)$$

In this case the *selectivity* is that fraction of the electrophile which is converted to the desired product (eqn. 4.13). The *selectivity* is determined by the rate of reaction to product divided by the sum of this plus the rate of the competing process:

$$selectivity = \frac{r_p}{r_p + r_{solv}} = \frac{k_p[E][Nuc]}{k_p[E][Nuc] + k_{solv}[E]} \qquad (4.13)$$

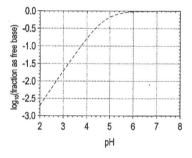

**Fig. 4.8** Plot of fraction of aniline present as free base *vs* pH (pK$_a$ of conjugate acid = 4.7)

Substituting eqns. 4.7 and 4.11 into 4.13, and using the transformation $[OH^-] = K_W/[H^+]$ gives eqn. 4.14.

$$\text{selectivity} = \cfrac{\cfrac{k_p[Nuc]_T}{\left(1+\cfrac{[H^+]}{K_N}\right)}}{\cfrac{k_p[Nuc]_T}{\left(1+\cfrac{[H^+]}{K_N}\right)} + \left(k_{H_2O} + \cfrac{k_{OH^-}K_W}{[H^+]}\right)} \qquad (4.14)$$

$$k_{solv} = k_{H_2O} + \frac{k_{OH^-}K_w}{[H^+]}$$

This expression is plotted in Fig 4.9, again in logarithmic form, for selected values of $k_{solv}$, $k_p$ and $[Nuc]_T$.

Data used for Fig. 4.7:
Electrophile, as for acetic anhydride (Table 4.4); nucleophile, assume $K_N = 10^{-5}$; $k_p = 15\,M^{-1}s^{-1}$, $[Nuc]_T = 0.1$.

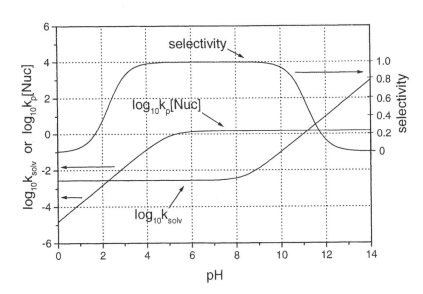

**Fig. 4.9** pH/rate profile for competing solvolysis and product formation: reaction of an electrophile with a base

Note that eqn 4.14 refers to the situation where the nucleophile is in excess over E, such that [Nuc] does not change during the process. The width of the plateau region where the selectivity is approximately constant is dependent on the difference between $pK_i$ and $pK_N$.

Four factors determine the behaviour of the system.

1.  The absolute value of $k_{H_2O}$ will determine whether a homogeneous process using water as solvent is feasible. Note also that the half lives for benzoyl chloride and for phosgene in water are less than one second, so that hydrolysis will occur rapidly before dissolution and mixing into the bulk solution can occur. Special techniques involving two-phase reactions have to be used to obtain satisfactory reactions with such reactive electrophiles.

2.  $k_{H_2O}$ and $k_{OH^-}$ together determine $pK_i$, the point at which the solvolysis switches from being pH independent to directly proportional to [OH⁻]. This is exemplified in Fig. 4.7.
3.  $pK_N$ determines the pH at which the free base is fully available, and together with $pK_i$ determines the width of the pH window where the selectivity is at a maximum. In the example shown the selectivity is approximately constant from pH 4 to 9.
4.  $[Nuc]_T$ , the concentration of the nucleophile, is a proportionality constant in the rate of the product forming step. Increasing the concentration will increase the selectivity.

When there is more than one reactive group in the same nucleophilic species the same principles can be applied to derive the optimum conditions required to react at one or other nucleophilic site. The case of a diamine with aromatic and aliphatic amino groups of different basicity is common, and is exemplified by the case of 4-aminobenzylamine (Fig. 4.10). By plotting the two ionisation curves together the behaviour of the system is easily seen. To a first approximation it can be assumed that eqn. 4.10 can be applied separately to each amino group. The region where discrimination is obtained on the basis of pH can be seen in Fig. 4.10.

In this case, if monoacylation of the aromatic amine is required, this will be achieved by working at the lowest pH where the aromatic amine is in the free base form and the more basic aliphatic amine is substantially fully protonated. By inspection of the plots it is evident that a pH of around 4 is optimum. It is not possible to achieve a selectivity towards monoacylation of the *aliphatic* amine better than provided by the relative reactivity of the two amino groups, which will be obtained when the aliphatic amino group is present as the free base. Optimum conditions in order to minimise electrophile hydrolysis will be the minimum pH necessary to ensure that the more basic amine is present in the free amine form, in this case around pH 11.

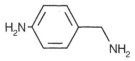

4-aminobenzylamine
$pK_a$ aromatic = 3.74
$pK_a$ aliphatic = 9.70

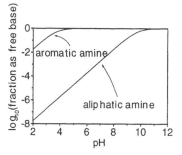

**Fig. 4.10** Differentiation by protonation of the two amino groups in 4-aminobenzylamine.

## 4.5 Reactor configuration and selectivity

The semi-batch reactor is extremely versatile and can cope with a variety of demanding chemistry. The order of addition can be chosen so that the exposure of reactants which are unstable under the reaction conditions *e.g* by solvolysis, can be minimised. Concurrent addition of two components can be employed where both are unstable, or where extended contact of a reactant with product is undesirable. Exothermic reactions can be controlled by addition of one component under conditions where the heat can be safely removed. Where pH control is required, due to the liberation of acid or base by the reaction, or because either is added with a reactant, then this is also possible. An appropriate buffer is required to avoid violent fluctuations in pH. Sometimes the hazardous nature of the process may lead to the selection of a continuous process option as a means of minimising the inventory of materials. Only in unusual circumstances is the use of alternative reactor configurations enforced by the chemistry. Two instances are:

*   Where the *product* is unstable on the timescale of the batch reaction, and
*   where rapid stoichiometric mixing of two components is required.

The first case applied to a process developed within Zeneca. Fig. 4.11 shows the synthesis.

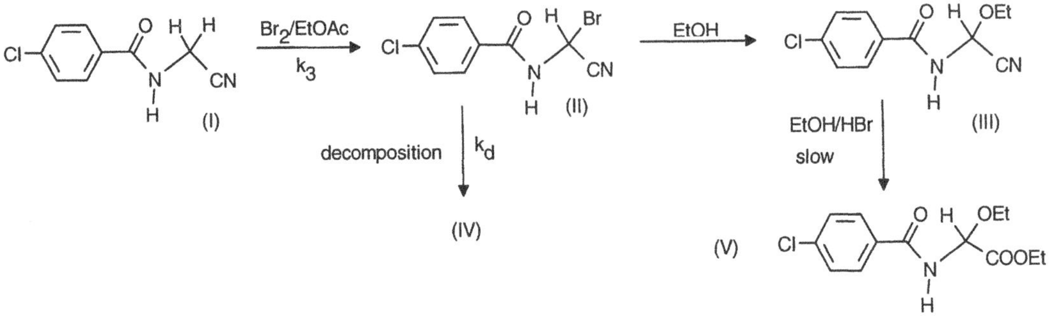

**Fig. 4.11** Synthesis of an α-ethoxynitrile.

On the laboratory scale it was easy to obtain a high yield. A stoichiometric quantity of bromine was added to a cold solution of the starting aminonitrile (I) in solution in ethyl acetate. Within five minutes the colour of bromine had disappeared, whereupon the solution was poured into ethanol, from which the product (III) precipitated in high yield. A more detailed examination of the chemistry suggested that there would be significant difficulty with scale-up in a batch or semi-batch reactor, for three reasons.

1    The bromination reaction is highly exothermic, leading to a calculated reaction time in a semi-batch reactor on the full scale, determined by heat transfer, of several hours.

2    Bromination liberates acid, is strongly acid catalysed and requires acid to proceed at a reasonable rate.

3    The intermediate bromo-compound (II) is very unstable under the acidic reaction conditions.

No permutation of batch processing options could be found which would avoid these difficulties. The most obvious solution, use of ethanol as solvent for the two stages, failed because the bromination rate was slowed down relative to reaction of the nitrile group with ethanol to give (V).

It was now apparent that to make (II) in good yield a short contact time reactor was required, and that backmixing of the product with the reactants would be the ideal way of providing the acid catalysis. A continuous stirred tank reactor (CSTR)(see Chapter 2) meets these requirements. The subsequent ethanolysis is simple and can be carried out in semi-batch mode. From the known rate constants and the chosen operating concentrations, the residence time necessary to maximise the yield can calculated.

The equations required to model the system are

$$I + Br_2 \xrightarrow{\text{HBr},k_3} II + HBr; \qquad \text{rate} = k_3[I][Br_2][HBr] \qquad (4.15)$$

$$II \xrightarrow{k_d,\text{HBr}} IV; \qquad \text{rate} = k_d[II] \qquad (4.16)$$

The procedure is to carry out a mass balance at steady state (Fig. 4.12).

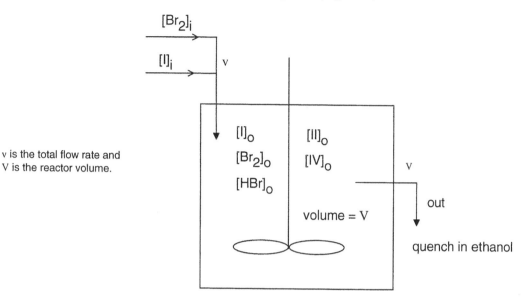

v is the total flow rate and
V is the reactor volume.

**Fig. 4.12** Reactor scheme for continuous bromination

A mass balance on (I) gives

$$v[I]_i = k_3 V[I]_o [Br_2]_o [HBr]_o + v[I]_o \qquad (4.17)$$

in              reacted        out

This rearranges to (4.18), where $t_r$ is the residence time, V/v.

$$\frac{[I]_o}{[I]_i} = \frac{1}{1 + k_3 t_r [Br_2]_o [HBr]_o} \qquad (4.18)$$

Mass balancing II gives

$$k_3 V[I]_o [Br_2][HBr]_o = k_d V[II]_o + v[II]_o \qquad (4.19)$$

which rearranges to

$$\frac{[II]_o}{[I]_o} = \frac{k_3 t_r [HBr]_o [Br_2]_o}{1 + k_d t_r} \qquad (4.20)$$

Multiplying 4.20 by 4.18 gives

$$\frac{[II]_o}{[I]_i} = \left\{ \frac{1}{1 + k_d t_r} \right\} \left\{ 1 - \frac{1}{1 + k_3 t_r [HBr]_o [Br_2]_o} \right\} \qquad (4.21)$$

Note that $[I]_i - [I]_o$ is the amount of aminonitrile reacted. It is not the same as the amount of (II) produced, since there is some decomposition of (II).

The stoichiometry balances are

$$[Br_2]_o = [Br_2]_i - \{[I]_i - [I]_o\} \qquad \text{and} \qquad (4.22)$$

$$[HBr]_o = [I]_i - [I]_o \qquad (4.23)$$

since one mol of bromine reacts with one mol of (I) to give one mol of product and one mol of HBr.

It is easiest to do the calculation of residence time by choosing values of conversion in the desired range, and calculating the corresponding residence time using eqn 4.24, which is a rearranged form of eqn 4.18, simplified using eqn 4.23.

$$t_r = \frac{[I]_i - [I]_o}{k_3[I]_o[Br_2]_o[HBr]_o} = \frac{1}{k_3[I]_o[Br_2]_o} \qquad (4.24)$$

Solving these for the concentrations of the various species *vs* time gives the results shown in Table 4.5 and Fig. 4.13.

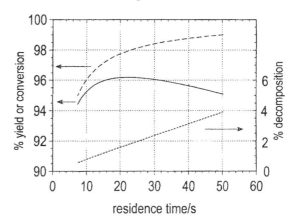

**Fig. 4.13** Conversion (dash), product decomposition (dot) and yield (full line) for the bromination of (I).

There is a calculated yield maximum of 96.2% at a residence time of 20–25 s. Higher *conversions* can be achieved by increasing the residence time, but then the amount of *decomposition* is increased.

**Table 4.5** Calculated reactor performance vs residence time

| Conversion %* | $[Br_2]_o$ | $[HBr]_o$ | $[I]_o$ | % Yield = $100[II]_o/[I]_i$ | % Decomposition = conversion-yield | $t_r$/s |
|---|---|---|---|---|---|---|
| 0 | 0.26 | 0 | 1 | 0 | 0 | 0 |
| 90 | 0.0476 | 0.2124 | 0.0236 | 89.8 | 0.2 | 2.8 |
| 95 | 0.0358 | 0.2242 | 0.0118 | 94.4 | 0.6 | 7.4 |
| 98 | 0.02872 | 0.2313 | 0.00472 | 96.2 | 1.8 | 23.1 |
| 99 | 0.02636 | 0.2336 | 0.00236 | 95.1 | 3.9 | 50.2 |
| 100 | 0.024 | 0.236 | 0 | | | $\rightarrow \infty$ |

*Conversion = $100 \times ([I]_i - [I]_o)/[I]_i$
$k_3 = 320$ M$^{-2}$ s$^{-1}$; $k_d = 8.2 \times 10^{-4}$ s$^{-1}$ in 0.236M HBr; both at 25ºC.

In the second example, from a Merck process, shown in Fig. 4.14, acylation of L-proline (LP) by L-alanine-N-carboxyanhydride (ANH) was complicated by further acylation of the desired product (P) to give the dipeptide (DP)

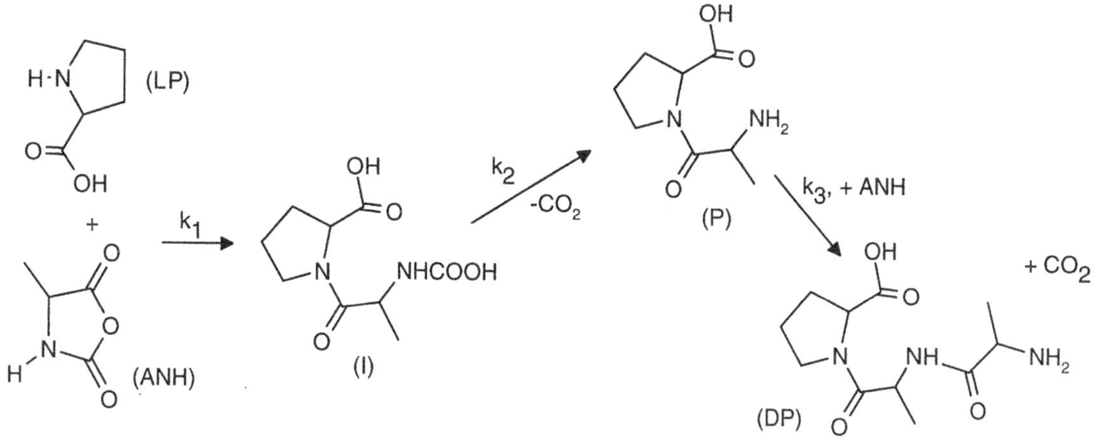

**Fig. 4.14** Competitive reaction in the acylation o L-proline.

It was found that, on the laboratory scale, the yield of product was dependent on the addition time of the anhydride to LP. Short addition times were best; extension of the addition time to 1000 seconds resulted in a yield loss of around 10%. This was attributed to formation of an unstable intermediate, probably the carbamic acid (I), which was much less reactive with anhydride than product (the amino group is protected). At short addition times the added anhydride acylates LP without competition with product, because the product is transiently protected as (I). At long addition times the protecting group is lost and competitive acylation occurs, so that the yield approximates to that determined by the ratio $k_1/k_3$ (see section 4.3).

Since very rapid addition times were implausible in a semi-batch reactor, the reaction was conducted in a tubular reactor (Fig. 2.4) with a residence time of around 1 second. This reactor permitted yields at full scale operation equal to those achieved in the laboratory.

$k_1 \approx 100$ M$^{-1}$s$^{-1}$
$k_2$ is unknown
$k_3 \approx 10$ M$^{-1}$s$^{-1}$
From Paul E.L. (1988). Design of Reaction Systems for Specialty Organic Chemicals. Chemical Engineering Science, 43 (8), 1773–1782.

## 4.6 Methods for predicting reactivity

Vast amounts of rate data are scattered throughout the literature; as with synthetic work, a detailed search can unearth much time-saving information. If no reference to the reaction of interest can be found, it is often possible, by careful choice of analogous systems, to make a close estimate of the required rate constant. Several methods are available for interpolating (or with less certainty, extrapolating) rate constants, which in appropriate circumstances can eliminate the need to carry out the measurements. These methods have been developed by chemists interested in elucidating the finer points of reaction mechanisms; they can also be extremely valuable to the practical chemist seeking to obtain an estimate of a rate constant without the trouble (and expense) of measuring it.

The methods use a relationship of the general form 4.22.

$$\log_{10}\left(\frac{k}{k_0}\right) = mx \qquad (4.22)$$

where $k_0$ is the rate constant for reaction of the reference compound (or solvent), $k$ is the rate constant for reaction of the unknown, $m$ is a slope term which is a constant for the reaction in question and $x$ is specific to the particular reactant or substituent. This expression can be written as 4.23.

$$\log_{10} k = mx + c, \quad \text{where} \quad c = \log_{10} k_0 \qquad (4.23)$$

Thus in the straightforward cases a straight line relationship is found between nucleophilic reactivity and a parameter related to the nucleophile. Some of the many useful correlations are shown in Table 4.6.

**Table 4.6** Some structure reactivity correlations.

| Name | Reaction type/examples |
| --- | --- |
| Hammett | Substituent effects on reactivity at aromatic substituents. $ArNMe_2 + MeI \rightarrow ArNMe_3^+ + I^-$ |
| Taft | Substituent additivity effects in aliphatic systems. Formate ester hydrolysis |
| Extended Bronsted | Reactivity/basicity correlation Acyl group transfer |
| Swain-Scott | Nucleophile/electrophile reactivity scales in protic solvents Rates of reaction of simple inorganic nucleophiles with esters. |
| Mayr | Nucleophile/electrophile reactivity scales in aprotic solvents (mainly dichloromethane.acetonitrile) $Ar_2CH^+$ with alkenes |
| Winstein-Grunwald | Solvent effects on $S_N1$ reaction rates. Ionisation of alkyl halides vs solvent polarity. |

One example will be discussed in more detail. The original Bronsted relationship connected the rate of transfer of a proton to a base with the $pK_a$ of the base. For a large number of electrophile/nucleophile reactions there is a linear relationship between $pK_a$ of a series of related nucleophiles and the reaction rate with a particular electrophile. This is referred to as the *Extended Bronsted Relationship* and may be written as 4.24.

$$\log_{10} k_N = \beta_N pK_N + C \qquad (4.24)$$

where $k_N$ is the rate constant, $\beta_N$ is a term characteristic of the sensitivity of the reaction rate to the basicity of the nucleophile, and $pK_N$ is the $pK_a$ of the nucleophile.

For the reaction of amines with a particular electrophile the plot of rate *vs* $pK_N$ should be a straight line with slope $\beta_N$, if attack of the amine is rate determining. An example is the hydrolysis of acetic anhydride catalysed by various pyridines (Fig. 4.15).

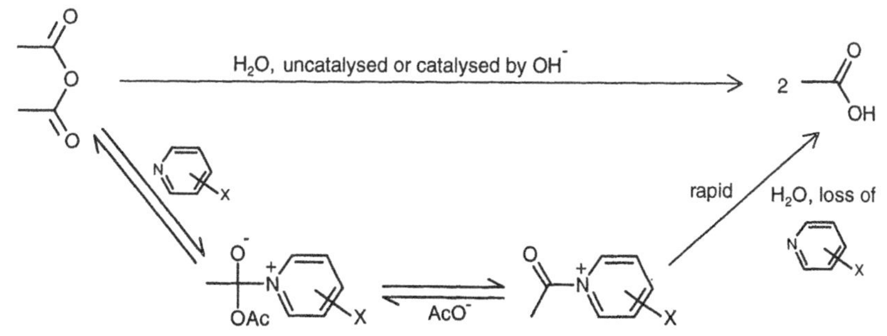

**Fig. 4.15** Catalysed hydrolysis of acetic anhydride.

Fig. 4.16 shows the data for the pyridine catalysed component of the hydrolysis of acetic anhydride. For pyridines with a $pK_a > 6.1$ the rate determining step is loss of the leaving group (acetate) from the tetrahedral intermediate, for which the slope of the plot of $pK_a$ vs $log_{10}k_N$ is 0.2. For pyridines with a $pK_a < 6.1$ the rate determining step is formation of the intermediate, for which the slope is unity. Interpolation to obtain the reaction rates for other pyridines is probably secure, but these data also illustrate the dangers of extrapolation without a knowledge of the mechanism.

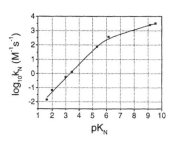

**Fig. 4.16** Plot of $log_{10}k_N$ *vs* $pK_N$ for the catalysed solvolysis of acetic anhydride.
Redrawn with permission from Castro C.and Castro E.A. (1981). J. Org. Chem., 46, 2939.

# Bibliography

Russell G.A. (1961). Competing reactions, *in Technique in organic chemistry, Vol. VIII-Part I*, Interscience.

Jencks W.P. (1987). *Catalysis in chemistry and enzymology*, Dover Publications.

Page M.I. and Williams A. (1997), *Organic and bio-organic mechanisms*, Longman.

H Maskill (1985), *The physical basis of organic chemistry*, Oxford University Press.

United States Department of Commerce (1951). Tables of chemical kinetics (Homogeneous reactions), in *Circular of the National Bureau of Standards no. 510.*

# 5. Mixing effects in pseudo - homogeneous systems

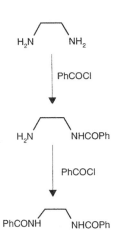

**Fig.5.1** Benzoylation of ethylene diamine at –78°C; Sayre J. (1987). J. Org. Chem., 52, 2592. At room temperature the rate constant for reaction of benzoyl chloride with aliphatic amines is about $10^4$ $M^{-1}$ $s^{-1}$ King J. (1992) J. Amer. Chem. Soc., 114, 3028; there are no data at –78°C., but the reaction is still expected to be rapid.

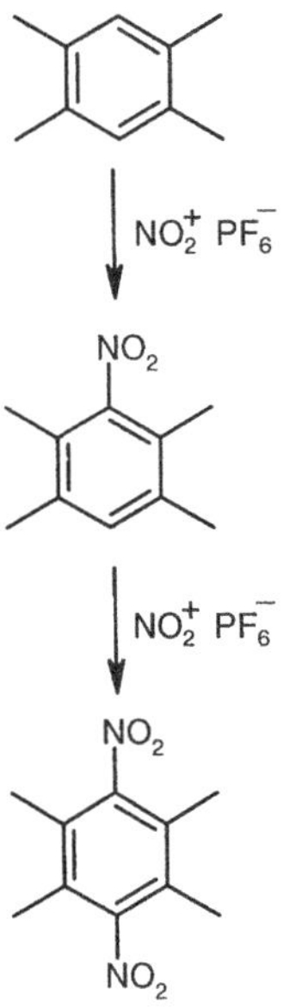

When competing reactions are very fast, unexpected and sometimes counter-intuitive selectivities can be obtained. Two simple examples illustrate this well. In Fig. 5.1, the benzoylation of ethylene diamine using a five-fold excess of the diamine, is illustrated. The reactivity of the amine groups is expected to be the same and each independent of the other half of the molecule, so that under these conditions a large excess of monobenzoyl compound would be expected. Instead, the experiment yielded largely the dibenzoyl compound.

**Table 5.1** Benzoylation of ethylene diamine with benzoyl chloride at –78°C. (%yield based on benzoyl chloride, 5 mol excess of amine)

| Conc. of diamine solution M | Conc. of PhCOCl solution M | % dibenzoylamine |
|---|---|---|
| 0.67 | 0.4 | 99 |
| 0.033 | 0.02 | 22 |

The explanation for the unexpected selectivity lies in the local inhomogeneity and the high reactivity of the system. On the rapid time scale of the competing second benzoylation, the mono- compound cannot mix with the bulk away from the benzoyl chloride and fresh diamine cannot be transported in to mix with the benzoyl chloride quickly enough to prevent the second benzoylation from occurring.

A second example is analysed in more detail in section 5.2. It is the nitration of tetramethyl benzenes, with equimolar nitrating agent, nitronium hexafluorophosphate, in nitromethane (Fig. 5.2). In this example the ratio of rate constants predicts:

mononitrodurene : dinitrodurene >1000 : 1

In practice the experiment yielded:

mononitrodurene : dinitrodurene = 1 : 13.5.

These examples illustrate that even in single phase reacting systems, local inhomogeneity can have a serious effect on selectivity when there are rapid competing by-product reactions. In semi-batch operation, where one or more reagent is added to a reaction mass, and in continuous reactors where reagents may be mixed in-line or in stirred tank reactors, basic reaction engineering theories assume the reactor contents to be homogeneous. However on length scales local to the addition point and time scales within the reactor residence time or reagent addition time, there will be regions of reactant concentration significantly different from the bulk average. The same can be true for gradients in temperature and pH. A simple comparison of the time scales on which these transients exist, with a characteristic reaction time

**Fig. 5.2** Nitration of tetramethylbenzene Rys P (1975). Helv. Chim. Acta, 58.

constant for the by-product reaction, gives an indication of the potential difficulty and allows problems to be classified for theoretical and experimental analysis .

For a simple second order reaction (rate constant $k$) the half life is a convenient measure of the time constant ($\tau_R$). For equal initial concentrations ($C_{A0}$), it is:

$$\tau_R = \frac{1}{kC_{A_0}} \tag{5.1}$$

For semi-batch operation in a stirred reactor it is more realistic to use the theoretical concentration which would exist after addition if there were no reaction:

$$\tau_R = \frac{1+a}{kC_{A_0}} \; where; a = \frac{V_A}{V_B} \tag{5.2}$$

$V_A$ and $V_B$ are the volumes of the reactants added.

The mixing regimes are classified as: Macromixing, Mesomixing or Micromixing,.

**Table 5.2** Mixin g regimes

| Reaction time constant $\tau_R$) | Mixing regime |
|---|---|
| 10 mins. – hrs. | independent of mixing |
| secs – mins | macromixing |
| ms | micromixing |

Mesomixing falls between macro- and micromixing. The names reflect the different length and time scales of the dominating mechanisms. *The macro scale relates to the scale of the equipment*, cm in the laboraory to several m at full scale. The micro scale relates to the size of the smallest turbulent eddies, which are µm regardless of the size of the equipment.

In order to be able to predict the effects of mixing on reaction selectivity during process development, it is necessary to appreciate  a number of concepts and definitions used to characterise fluid flows. In tubular flow and in agitated vessels, the concepts of turbulent, laminar and transitional flows are important. *In order for a vessel to be well mixed at any length scale it would normally have to be operating under turbulent conditions.* Turbulence is achieved when inertial forces dominate over viscous forces. The ratio of these forces is the well known *Reynolds Number,* denoted by the symbol Re and defined in equations (5.3) and (5.4). At high values (Re>3500 in a tube, Re>1000 in an agitated vessel) the flow is turbulent, at low values (Re<1000 in a tube and Re<10 in a vessel) it is laminar. At intermediate values, the flow is transitional, between the two main flow regimes.

Laminar flow occurs when the fluid flow pattern is well defined, the fluid moves in streamlines with no movement lateral  to the flow direction. Turbulent flow occurs when there is random motion imposed upon the main flow direction, with fluctuating motion in all directions. The fluctuations exist as whirling vortex  structures, or eddies, with dimensions from the scale of the equipment down to very small,  micron scales.

The Reynolds number for flow in tubular reactors is based on the tube diameter ($d$) and mean velocity ($u$) , with the fluid physical properties, bulk

An example calculation of the flow and tube dimensions required to achieve turbulence in a tubular reactor is presented in section 5.1.

density ($\rho$) and dynamic viscosity ($\mu$). It is a ratio of forces and is therefore dimensionless, but requires consistent units.

*Flows with the same Reynolds number have the same turbulence properties even when the velocities, pipe diameters or physical properties are different.*

*The Reynolds number determines whether the flow is laminar and therefore very poorly mixed; or turbulent.*

$$Re = \frac{du\rho}{\mu} \tag{5.3}$$

For agitated vessels the *impeller Reynolds Number* is used, based on the impeller diameter ($D$) and tip speed ($\propto ND$, where ($N$) is the rotational speed):

$$Re = \frac{ND^2\rho}{\mu} \tag{5.4}$$

**SI units are used throughout:**

| | |
|---|---|
| Density ($\rho$) | kg m$^{-3}$ |
| Dynamic viscosity ($\mu$) | N s m$^{-2}$ |
| Energy dissipation ($\varepsilon$) | W kg$^{-1}$ |
| Friction factor *(f)* | $-$a |
| Flow rate *(Q)* | m$^3$ s$^{-1}$ |
| Kinematic viscosity *(v)* | m$^2$ s$^{-1}$ |
| Length *(D, d, L, z)* | m |
| Power *(P)* | W |
| Power number *(Po)* | $-$a |
| Pressure *(p)* | N m$^{-2}$ |
| Rotational speed *(N)* | s |
| Time *(t)* | s |
| Volume *(V)* | m$^3$ |

aDimensionless.

It describes the flow in the immediate vicinity of the agitator. For low viscosity liquids turbulent flow around the agitator is a good indicator of turbulence throughout the vessel. This is not necessarily the case for a high viscosity or non-Newtonian fluid. A non-Newtonian fluid is one where the viscosity is not constant under all flow conditions, that is the shear stress is not proportional to the shear rate imposed on the fluid. In practise there are many forms of behaviour, but the most common in fine chemicals manufacture is when the fluid behaves as a very viscous liquid or even a gel under low shear rates, but like a low viscosity fluid at higher shear rates. It is therefore possible to have turbulent flow around an agitator, but stagnant regions near the vessel walls.

The energy present in the flow is an important parameter. In tubular reactors it is the pressure, derived from the pressure drop. The pressure drop is made up of the kinetic energy lost due to friction at the walls and the potential energy due to the difference in static head according to:

$$\Delta p = f\frac{L}{d}\rho u^2 + \rho g(z_1 - z_2) \tag{5.5}$$

$f$ is the friction factor which, in turbulent flow, depends on the pipe roughness. $L$ is the reactor length, and $(z_1 - z_2)$ the difference in elevation between inlet and outlet.

For an agitator the energy used to create the flow is the agitator power ($P$), calculated from the drag exerted by the fluid on the agitator:

$$P = P_o N^3 D^5 \rho \tag{5.6}$$

$P_o$ is the power number which, like $f$ above, depends upon Re, as well as the agitator and vessel design. Typical values range from 0.3 for efficient turbines which generate high flows, with little local turbulence, to 6 for agitators designed to input high levels of turbulent energy and lower flows.

*$\varepsilon$ is used to determine micromixing and mass transfer rates*

An important parameter in predicting micromixing in turbulent flows is the energy dissipation rate ($\varepsilon$). It is also used to predict mass transfer rates in multi - phase systems (Chapter 7).

For an agitated vessel: $\varepsilon_{mean} = \dfrac{P}{V\rho}$ , for a tubular reactor: $\varepsilon_{mean} = \dfrac{Q\Delta p}{V\rho}$

where $Q$ is the flow rate and $V$ the reactor volume. In the following sections these parameters will be used to characterise mixing regimes in reactors.

## 5.1 Macromixing

A reaction where the selectivity is affected by concentration gradients on the length scale of the equipment, is said to be dominated by macromixing . This can be from $10^{-3}$m at laboratory scale and up to 10m at full scale. If the reactor geometry and fluid flow field are simple and the number of competing reactions small and easily described, then it may be possible to predict selectivity from an understanding of the reaction mechanism and kinetics with a mathematical description of the flow. For most reactors, however, and particularly for agitated vessels commonly used in batch chemical manufacture, the flow field is extremely complex. It is three dimensional and time dependent. Recent developments in computational fluid dynamics have led to the development of time averaged approximated models of the fluid mechanics, including simple reaction schemes. As yet it is not realistic to expect to be able to use them routinely to predict the selectivity of complex reaction schemes in stirred batch reactors. However, as shown in the following sections, *it is possible to use a small laboratory scale model reactor to predict the full scale selectivity of a macromixing dominated reaction using the concept of Mixing Time.*

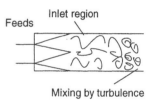

Fed by turbulence

**Fig. 5.3** Mixing in a tubular reactor

### Macromixing in tubular reactors

For tubular reactors in turbulent flow, the mixing process is normally represented as illustrated in Fig. 5.3. There is an inlet region where the feeds are bulk mixed by macro-scale distribution and dispersion determined by the inlet conditions and bulk flow properties, followed by diffusive mixing. At realistic flows a tubular reactor has a short residence time and is therefore only suitable for rapid reactions. The reactions are so rapid that the selectivity is insensitive to the macro-scale inlet conditions, but may still be influenced by meso- or micro-mixing.

In order to achieve a desired degree of conversion, it is necessary to achieve a specified and controlled residence time. A controlled residence time is obtained by generating a "plug flow" velocity profile, where the fluid flows with the same velocity across the diameter of the reactor tube, like a solid plug. In this way a flat concentration profile across the tube radius at any point, is achieved. To ensure "plug flow", fully turbulent flow is required, which, as described previously means that the velocity and reactor diameter must be set to achieve a Reynold's Number of at least 3500. This will require a high velocity, therefore *tubular reactors can only be used for fast reactions and large instantaneous production rates.* Static mixers or inserts inside the tube can be used to extend the applicability to a limited extent. However, for longer reaction times mixing has to be provided by mechanical means, normally in agitated vessels. They can be used in batch mode and individually or in cascades for continuous or semi-continuous

**An example of the flow conditions required to ensure turbulence in water:**
The density and viscosity of water are such that:

$\dfrac{\rho}{\mu} = 10^{6}$ m$^2$s$^{-1}$

The required Reynolds number is:

Re= $\dfrac{du\rho}{\mu} = 3500$

therefore:

$du = 3.5*10^{-3}$ m$^2$s$^{-1}$

Hence, for a 25mm tube

$u = 0.14ms^{-1}$

and the flow rate is 0.2 m$^3$hr$^{-1}$ or 4.8 m$^3$day$^{-1}$.

For a 10m long reactor, at this velocity, the maximum residence time would be 71s operation (*CSTRs*). It is difficult to operate a laboratory scale tubular reactor with a well controlled residence time without using excessive quantities of materials as the small diameter means that a very high flow rate is required to achieve turbulence.

## Macromixing in agitated vessels - flow patterns and agitator design

Most laboratory reactors with a rapidly moving agitator and a low viscosity mixture appear well mixed. For higher viscosity and non-Newtonian mixtures the degree of mixing becomes poor as the viscosity increases until a judgement will be made that it can no longer be handled. In a full scale reactor, even at low viscosity the design of the agitator and reactor internal fittings are important to achieve the degree of mixing the reaction requires. At higher viscosity the materials can be handled more readily than in the laboratory, but with more complex design of equipment. In full scale reactors, agitators are normally small diameter high speed turbines (Fig. 5.4), or larger diameter close clearance, that is the gap between the outer edge of the agitator and the vessel wall is small (Fig 5.6).

The small diameter turbines should only ever be operated without baffles in very simple single phase low viscosity mixtures. *For all other cases baffles are essential for good dispersion, especially in two phase systems* (Chapter 7). The baffles convert circumferential motion into vertical flow, generating a good flow pattern for mixing . Turbines like the top example in Fig 5.4, can provide a completely axial flow, discharging vertically downwards from the agitator, sweeping the vessel base and returning via the surface (Fig. 5.5). These are good for low viscosity mixing and suspending solid particles (Chapter 7). Those like the centre example, a disc turbine, provide radial flow, discharging towards the walls, with the vertical motion above and below the agitator plane generated by the baffles (Fig. 5.5). They are good for producing a lot of energy local to the agitator for mass transfer between liquids or liquids and gases. The bottom example in Fig. 5.4 is an angled blade turbine, which provides a general purpose mixed flow and can be used for many duties.

Close clearance agitators, particularly the commonly used anchor (Fig 5.6) are really only suitable for high viscosity duties, especially where heat transfer to the wall is required.

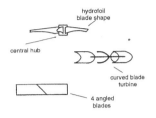

**Fig. 5.4** Three examples of turbine agitators (side views, shaft not shown)

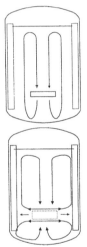

**Fig. 5.5** Axial and Radial flow patterns

## Macromixing in agitated vessels - mixing time

A reactor mixing time is commonly defined as the time taken to blend an inert tracer component into a liquid of the same physical properties, for example a dilute salt solution into water, to an end point. The end-point generally adopted as an industrial standard for agitated vessels is the time at which the variance in concentration measured at a number of points falls to within a defined fraction of the initial tracer concentration. For a variance of 5% of the initial concentration, the time would be $t_{95}$ and the reactor said to be 95% mixed. A similar approach can be used for the mixing length at the inlet of tubular reactors.

   Mixing time is a rather arbitrary parameter, but is defined in such a precise way to allow comparison of one vessel with another on a consistent basis. Reactors with the same mixing time will have the same macromixing behaviour *regardless of their size*. A consistent definition also allows mixing time to be correlated with system variables and physical properties so that it can be calculated for any reactor without having to be measured. In this way *the performance of a full scale reactor for a complex system of competing reactions can be modelled in the laboratory by using a small scale reactor with the same mixing time* without having to measure it. Similarly mixing time can be used as a design parameter. For example if a reaction is developed in the laboratory and it is determined that satisfactory selectivity can be obtained in a vessel with a mixing time of 10s, then a full scale reactor can be designed to achieve a 10s mixing time. It is more difficult to achieve a short mixing time at large scale because the power required to drive the agitator can become unrealistic. Nevertheless, even for very large reactors, for example 60m$^3$, a $t_{95}$ of 20 – 40s is easily achievable with the correct agitator and baffle design for a low viscosity reaction mixture and would be regarded as reasonably well mixed on a macro-scale.

   There is a critical Reynold's number (Re$_c$), above which the bulk mixing is turbulent and independent of the actual value of Re. It can be calculated using the agitator power number *(Po, described previously)* and an empirical constant measured for a range of agitators over a wide range of scales:

$$Re_c = 6370/Po^{\frac{1}{3}} \qquad (5.7)$$

For Re above the critical value, the following semi-empirical equation applies:

$$Po^{\frac{1}{3}} Re\, Fo = 5.2 \qquad (5.8)$$

*Fo* is a dimensionless group called *the Fourier Number*; $Fo = \dfrac{\mu t_{95}}{\rho T^2}$ (*T* is the reactor diameter (m)).

   This correlation can be used to calculate the mixing time for any agitated vessel reactor. It is possible to set up a model reactor at small scale with the linear dimensions all in the same proportion to the full scale, this is called geometrical similarity. For geometrical similarity, *Po* and *D/T* are constant. Therefore for a geometrically similar model reactor:

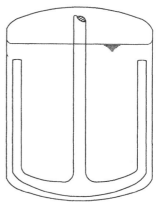

**Fig. 5.6** Close clearance Anchor agitator

*A laboratory scale reactor can be used to predict the performance of a full scale reactor by operating at the same mixing time*

**An example of a mixing time calculation for a full size industrial reactor:**
(Summarised in Carpenter K.J., (1997), in "A Handbook of Batch Process Design", Ed. Sharratt P.N., Blackie)

For a 20m$^3$ reactor, diameter *T = 3m*, with an angled blade agitator diameter *D = 1.5m* and *Po = 1*, operating at a rotational speed *N = 1s$^{-1}$* in water:
$\rho = 10^3\ kgm^{-3}$; $\mu = 10^{-3}\ Nsm^{-2}$:

$$Re = \frac{ND^2\rho}{\mu} = 2.25*10^6 >> Re_c$$

$$Fo = \frac{5.2}{Po^{\frac{1}{3}} Re} = 2.31*10^{-6}$$

$$t_{95} = Fo\frac{\rho T^2}{\mu} = 21s$$

The power input is:
$$P = PoN^3D^5\rho = 7600W = 7.6kW$$

and: $\varepsilon_{mean} = 0.38Wkg^{-1}$

**An example of the calculation of the minimum scale for a laboratory simulation of a full scale reactor:**

Consider an example where the reactor has full baffles and an agitator whose *Po* value is 1 ; with a water - like fluid, therefore:

$\rho/\mu = 10^6 \ m^2 s^{-1}$

In order to model a large scale reactor at small scale using a geometrically similar vessel, all geometrical ratios will be constant, therefore the small scale *Po* is also 1. Thus (from equation 5.7):

$ND^2 \geq 6370 * 10^{-6}$

The same mixing time is achieved at the same rotational speed regardless of scale for geometrical similarity, hence if at full scale:

*N=1 s$^{-1}$*, then *D>0.08m or 8cm.* Thus the agitator must be greater than 8cm diameter.

$$\text{Re } Fo = const. \text{ i.e } \frac{ND^2 \rho}{\mu} \frac{\mu t_{95}}{\rho T^2} = const. \text{ therefore } Nt_{95} = const.$$

This means that provided Re>Re$_c$ and strict geometrical similarity is maintained, *then the same mixing time or macromixing performance is obtained at the same agitator rotational speed, regardless of scale.*

The requirement for Re>Re$_c$ also provides a way of determining the minimum scale of vessel which can be used reliably to model a full scale reactor.

## 5.2  Micromixing

Micromixing  problems are caused by local inhomogeneity on a length scale at the opposite end of the turbulence spectrum to macromixing.  In this case the length scale is the smallest at which the flow still has turbulent characteristics and is, for example, around 30μm in turbulent water. The timescales for both the product reactions and the competing reactions which cause the loss in yield are of the order of ms. The phenomenon was described for the nitration of tetramethyl benzene with equimolar nitrating agent, nitronium hexafluorophosphate in nitromethane in Fig.5.2. The ratio of the rate constants for the  mono - and di- nitration reactions predicts a selectivity ratio of:

$$\frac{k_1}{k_2} \geq 10^3 \quad ; \text{ hence } \left\{ \frac{[mononitro]}{[dinitro]} \right\}_{theory} \geq 1000$$

However, on carrying out the reaction, the observed selectivity is actually in the opposite direction:

$$\left\{ \frac{[mononitro]}{[dinitro]} \right\}_{observed} = \frac{1}{13.55}$$

The original hypothesis was that although the second nitration reaction is slower than the first due to deactivation by the first nitro- group, it is still extremely fast. It is so fast that it occurs before the mononitro- compound has been able to diffuse away from the nitrating agent surrounding it, in which case, the selectivity is determined not by the ratio of rate constants, but by the ratio of the by-product reaction rate to the *diffusion rate*.

This simple theory does not fit experimental observations well and a more complex fluid dynamic model of the local mixing process than simple molecular diffusion is required. The most well recognised model is used here.

*Baldyga and Bourne Micromixing Model*

It can be assumed that negligible reaction occurs until the stream of added reagent has been broken down to the size of the most energetic eddies in turbulent flow, where there will be the greatest contact between the local regions of reagents. The theory can be applied to both agitated vessels and tubular reactors.  Analysis of the fluid dynamics shows the size of the most energetic eddies to be 12λ$_K$ where λ$_K$ is the Kolmogoroff microscale. It is the smallest size of eddy in the flow where the fluid is still turbulent. In eddy structures smaller than this, the fluid is in laminar flow and therefore mixes

(Baldyga and Bourne, Chem. Eng. J., 42,1989)
(Baldyga and Bourne, Chem. Eng. Commun., 28, 1984):

only by molecular diffusion. The Kolmogoroff microscale can be calculated from the physical properties and the energy in the flow:

$$\lambda_K = \left\{ \frac{v^3}{\varepsilon} \right\}^{\frac{1}{4}}$$
(5.9)

$\lambda_K$ is the smallest scale of a turbulent eddy, $\varepsilon$ is the energy dissipation rate and $v$ is the kinematic viscosity ($\mu/\rho$). Eddies of this size decay and their energy is dissipated as heat. The lifetime of such an eddy is $\tau_K$ where:

$$\tau_K = 12 \left\{ \frac{v}{\varepsilon} \right\}^{\frac{1}{2}}$$
(5.10)

It is postulated that during its lifetime, the eddy entrains fluid from its surroundings and becomes engulfed by an equal volume causing mixing and reaction. The reaction zone therefore grows due to this mixing or engulfment process. The rate of increase in the reaction zone volume $V_r$ can be represented by:

$$\frac{dV_r}{dt} = E V_r$$
(5.11)

$E$ is the engulfment rate and can be related to the flow properties and energy:

$$E = \frac{\ln 2}{\tau_K} = 0.05776 \left\{ \frac{\varepsilon}{v} \right\}^{\frac{1}{2}}$$
(5.12)

For low viscosity reaction media, it can be shown that it is this engulfment process rather than molecular diffusion which determines the rate of mixing of the reagents, hence rate of reaction. The mass balance for the added reagent becomes:

$$\frac{d(V_r c_i)}{dt} = E V_r [c_i] + R_i V_r$$
(5.13)

*[c$_i$]* and $R_i$ represent the concentration and the reaction rate of species $i$. This equation coupled with the rate of change of reacting volume $V_r$ (equns. 5.11 and 5.12) can be solved to predict the concentrations of reaction products and selectivity throughout the reaction. It is important to recognise that the energy dissipation rate used in this analysis and which determines the selectivity, is that *local to the addition point. It is therefore most important that the reagents are added into a region of strong turbulence.*

*Experimental and design implications*
Without solving the reaction and engulfment rate differential equations, the basic theory can still be used to indicate when problems will arise and to design experiments to predict the extent of the problem.

**An example of micromixing in a full scale reactor containing a turbulent water-like fluid:**

For a 20m³ reactor with an agitator power input of 20 kW, $\varepsilon_{mean} = 1 Wkg^{-1}$

If the feeds are added into the agitator region, it can be assumed that $\varepsilon_{local} \sim 50^*\varepsilon_{mean} = 50 \, Wkg^{-1}$
By calculation (equations 5.9 and 5.10):
$\lambda_K = 12\mu m$ and $\tau_K = 1.7ms$
The engulfment rate is: *410 s⁻¹* and the time constant *2.4 ms*
The by-product reaction time constant which will cause a micromixing problem is $\tau_R < 2.4ms$, thus for example, for a 1M solution $k > 410 \, M^{-1}s^{-1}$; that is a by product reaction with a rate constant up to 410 M⁻¹ s⁻¹ can be handled without problems.

If the feeds are added away from the agitator, but in the bulk of the liquid, $\varepsilon_{local} \sim 0.2^*\varepsilon_{mean} = 0.2Wkg^{-1}$
$\lambda_K = 50 \, \mu m$ and $\tau_K = 27ms$
The engulfment rate is: *25 s⁻¹* and the time constant *40ms*
The by-product reaction time constant which will cause a micromixing problem if the feeds are added here is
$\tau_R < 40 \, ms$, i.e. for a 1M solution: $k > 25 \, M^{-1}s^{-1}$

A problem arises when the by-product reaction rate is much faster than the rate of engulfment. Defining a time constant for the engulfment process as:

$$\tau_E = \frac{1}{E} \tag{5.14}$$

Then a problem of selectivity being determined by micromixing rather than intrinsic kinetics will exist when:

$$\frac{\tau_E}{\tau_R} >> 1 \tag{5.15}$$

*Again note that $\tau_E$ is determined by the energy dissipation rate local to the feed point ($\varepsilon_{local}$).* Typical values for common mixing and reaction equipment are:

**Table 53**   Energy dissipation rate and micromixing times for common equipment

| Equipment | $\varepsilon$ (Wkg$^{-1}$) (to order of magnitude) | Micromixing time in water (ms) |
|---|---|---|
| Tubular reactor or centrifugal pump | 5 | 8 |
| Stirred tank reactor | 0.2 – 50 | 40 – 2.4 |
| Static in - line mixer (e.g. Kenics) | 1000 | 0.5 |
| Rotor - Stator mixer device (e.g. Silverson mixer) | 5000 | 0.25 |

For a known reaction, it is therefore possible to select appropriate equipment so as to move away from a micromixing dominated regime at a given reagent concentration. *Simple techniques for moving away from micromixing control include slowing the reactions by diluting the reagents or by manipulation of pre-equilibria (Chapter 3).* This analysis gives a simple means of estimating the change required for a given reactor type.

The analysis also provides a means of designing laboratory experiments to estimate micromixing dominated yields. The engulfment rate is determined by the local energy dissipation rate, which is determined by the energy input per unit mass. To a first approximation it can be assumed that points at geometrically similar positions at different scales will have the same local value for the same overall mean, therefore *for laboratory simulation, reactions should be carried out at the same mean energy / unit mass as intended for full scale*.

## 5.3   Mesomixing

Baldyga and Bourne (Chem. Eng. Sci., 47. 8, 1992)

When feeding a highly reactive reagent into a stirred vessel in semi-batch mode the reaction selectivity can be determined by the rate of feeding rather than the reaction kinetics alone. If the feed rate is fast enough a plume of reagent will exist at the exit of the feed pipe, persisting for some distance then gradually dissipating by turbulent dispersion. A good example is the competing reaction of NaOH fed into an equimolar mixture of HCl and ethyl monochloroacetate at a volume ratio of 1:50 and stoichiometric ratio 1:1:1:

$$NaOH + HCl \xrightarrow{k_1} NaCl + H_2O$$

$$NaOH + CH_2ClCOOC_2H_5 \xrightarrow{k_2} C_2H_5OH + CH_2ClCOONa$$

At 298K: $k_1 \approx 10^{11} \, M^{-1}s^{-1} and, k_2 = 31 M^{-1}s^{-1}$

The selectivity of the ester hydrolysis to the acid - base reaction is :

$$X_{EtOH} = \frac{C_{EtOH}}{(C_{NaCl} + C_{EtOH})} \qquad (5.16)$$

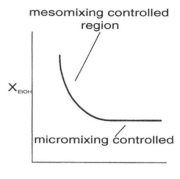

Fig. 5.7 Selectivity as a function of feed rate

Under these conditions, slow feeding results in the highest selectivity and the minimum amount of ethanol, as predicted from micromixing control described in section 5.2,. *Rapid feeding over a period less than a critical value, results in an undispersed plume of feed and poorer mixing* . The result is a local excess of NaOH, which produces more ethanol as shown in Fig. 5.7. This is the mesomixing controlled region.

Prediction of the effect of mesomixing requires knowledge of the turbulent diffusivity and velocity local to the feed point, parameters which are very system specific. Laboratory simulation of full scale operation is also not reliable in this region. *The details of the feed pipe, the turbulence characteristics of the feed and reactor contents at the exact feed point, and the feed rate are all important and it is not possible to match these exactly at small scale.* It is possible to predict to a first approximation whether a mesomixing problem will arise, simply by knowing if the time scale of the by-product reaction falls between micro- and macromixing times. If this is the case, and if it is not possible to move more towards the macro - scale by, for example slowing the reaction down, then further analysis is required, with a detailed knowledge of the reactor and feed arrangement to be used.

For further reading see :Baldyga et al (Chem. Eng. Sci. 48, 19, 1993)

# 6 Equilibria in multiphase systems

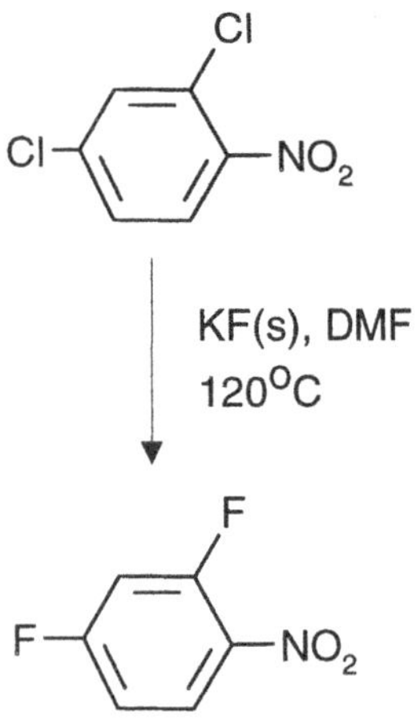

**Fig. 6.1** Halogen exchange in a heterogeneous reaction system (DMF = dimethylformamide)

Many processes are operated under conditions where more than one phase is present. This is often due to economic considerations: when it is necessary to contact large water-insoluble reactants with inorganic reagents, reactions are frequently run with either a separate solid phase or else a solution phase containing the inorganic reagent. An example, discussed in chapter 8, is the halide exchange reaction between a chloroaromatic compound and potassium fluoride, the latter having a very low solubility in the chosen solvent (dimethylformamide or dimethylacetamide) even at the reaction temperature of around 120°C (Fig. 6.1). A proper understanding of the process requires a knowledge of the solubility under reaction conditions.

Workup processes frequently involve washing of a solution of a water insoluble product which is dissolved in an organic solvent such as toluene with water in order to remove inorganic residues. These processes can involve numerous subtleties. Some of these are connected with the equilibrium aspects and are discussed here; dynamic aspects are covered in Chapter 7.

## 6.1 Simple distribution equilibria

**Table 6.1** Some $\log_{10}P$ values

| Compound | $\log_{10}P$ |
|---|---|
| morpholine | −1.08 |
| dimethyl ether | 0.1 |
| N-phenylacetamide | 1.16 |
| salicylic acid | 2.24 |
| cyclohexane | 3.44 |
| anthracene | 4.45 |

**Distribution of neutral organic species between water and solvent**

For the distribution of uncharged species between a solvent and water the partition coefficient $K_D$ is given simply by eqn 6.1.

$$K_D = \frac{c_{org}}{c_{aq}} \tag{6.1}$$

where $c_{org}$ and $c_{aq}$ refer to the concentrations in the organic and aqueous phases.

Vast compilations of partition coefficients are available. The octanol/water partition coefficient P, usually expressed in the logarithmic form as $\log_{10}P$, is used to estimate the absorption characteristics of pharmaceutical and agrochemical active ingredients into biosystems. Octanol is a convenient model for natural lipids. Some representative values are shown in Table 6.1.

$$P = \frac{c_{octanol}}{c_{aq}}$$

**Salt effects on solubilities of organic species**

The solubility of organic species in water is affected by the presence of dissolved salts. There is a simple but useful empirical correlation (eqn. 6.2), attributed to Setschenow, between the solubility of a non-electrolyte and the concentration of a dissolved salt.

Long F.A. and McDevit W.F. (1952). Chemical Reviews, 51, 119.

$$\log_{10}\left(\frac{S_0}{S_i}\right) = k_s C_s \tag{6.2}$$

In eqn. 6.2 $S_0$ is the solubility of the compound in water, $S_i$ is the solubility at electrolyte concentration $i$, $k_s$ is a constant specific to the particular substrate and electrolyte and $C_s$ is the electrolyte concentration in mol/l. Fig. 6.2 shows the effect of various electrolytes on the solubility of aniline in water, expressed as $\log_{10}(S_0/S_i)$.

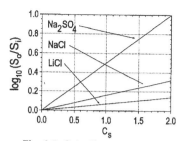

**Fig. 6.2** Salt effects on the solubility of aniline at 25°C.

### Phase-transfer catalysts

Phase transfer catalysis relies on the partitioning of a lipophilic pair of ions between an aqueous and an organic phase. In these simple cases the partitioning is described by eqn. 6.1. The partition coefficients vary in a logical way depending on the size of the organic ion, the degree of hydration of the inorganic ion and the organic solvent. Table 6.2 shows the effect of cation structure on the partition coefficient for some commercially available salts.

**Table 6.2** Partition coefficients for several quaternary ammonium chlorides between 1,2-dichloroethane and water at room temperature

| Cation | $K_D$ |
|---|---|
| $Bu_4N^+$ | 0.263 |
| $Bu_3N^+CH_3$ | $1.5 \times 10^{-3}$ |
| $(C_{8-10}H_{17-21})_3N^+CH_3$ | 323 |

Data from Halpern M. and Grinstein R. (1998). Choosing a phase-transfer catalyst to enhance reactivity and catalyst separation, in *Pilot plant and scale-up of chemical processes II*, Royal Society of Chemistry, London.

Not all anions are equally well partitioned into an organic solvent. Large lipophilic (oil-loving) anions are most easily transferred. Small hydrophilic anions such as fluoride, hydroxide and sulphate are hardly transferred at all. Table 6.3 shows some values for selectivity constants versus chloride as the reference anion. The selectivity constant $K_{Cl \to X}$ is defined (eqn 6.3) as

$$K_{Cl \to X} = \frac{[X^-]_{org}[Cl^-]_{aq}}{[X^-]_{aq}[Cl^-]_{org}} \qquad (6.3)$$

where the suffixes refer to the organic and aqueous phases.

The use of lipophilic leaving groups (bromide and iodide are two examples) can cause catalyst poisoning when these anions are transported in preference to the desired anion. The use of a chloro compound as a substrate would generally be preferred.

**Table 6.3** Some selectivity constants for extraction of various anions into toluene

| Anion | $K_{Cl \to X}$ |
|---|---|
| $OH^-$ | 0.01 |
| $F^-$ | 0.02 |
| $HCO_3^-$ | 0.05 |
| $Cl^-$ | 1 (reference) |
| $Br^-$ | 16.5 |
| $I^-$ | 5000 |
| $MnO_4^-$ | large |

Taken from: Starks C., Liotta C. and Halpern M. (1993). Phase transfer catalysis: fundamentals, applications and industrial perspectives, Chapman and Hall.

## 6.2 Solubilities of ionisable substrates

Work-up processes, because they involve a separation process of some sort - distillation, extraction or crystallisation - frequently involve ionic equilibria coupled to solubilities or partition coefficients. A simple example is the effect of pH on the solubility of an ionisable solid. Frequently the ionised form is more soluble in water than the neutral (free acid) form. In order to optimise the recovery from a crystallisation process it is necessary to know the solubility of product under the conditions of separation from the mother liquor. If a more soluble ionised form is present in equilibrium with the neutral species, then the solubility will be enhanced and the recovery efficiency reduced.

For an acidic species, ionisation of the free acid SH is described by the equilibrium shown in eqn. 6.4,

$$SH \xrightleftharpoons{K_a} S^- + H^+ \tag{6.4}$$

and the equilibrium constant by eqn. 6.5

$$K_a = \frac{[S^-][H^+]}{[SH]} \tag{6.5}$$

The total solubility $S_T$ at any pH is the sum of the solubility of the free acid SH, and the ionised form $[S^-]$ (eqn.6.4),

$$[S]_T = [SH] + [S^-] \tag{6.6}$$

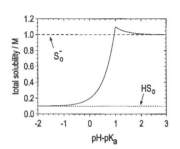

**Fig. 6.3** Total solubility of an acidic species vs pH for $HS_0$ = 0.1M; $S_0^-$ = 1.0M.

As the pH is increased from well below the $pK_a$ the limiting factor to start with is the solubility of the free acid, $SH_0$. Substituting eqn. 6.5 into 6.6 and simplifying gives 6.7,

$$[S]_T = [SH]_0 \left\{ 1 + \frac{[K_a]}{[H^+]} \right\} \quad \text{(acid solubility limited)} \tag{6.7}$$

which shows how the total solubility changes with the pH of the solution, up to the solubility limit of the anion. Above this pH the total solubility falls again, because now the amount of free acid falls with increasing pH. In this region the limiting solubility is that of the anion, $S_0^-$, so now the total solubility is given by

$[SH]_0$ is the solubility of the acid form; $[S^-]_0$ is the solubility of the anion form

$$[S]_T = [S^-]_0 \left\{ 1 + \frac{[H^+]}{[K_a]} \right\} \quad \text{(anion solubility limited)} \tag{6.8}$$

Fig. 6.3 shows how the solubility varies when the solubility of the acid is 0.1M and the anion 1M. The abscissa is normalised by subtracting the $pK_a$ from the pH of the solution. Either the actual or the logarithm of the concentration can be used depending on the concentration range involved.

These relationships are important in determining species solubilities in crystallisation processes, and also in determining the concentration of the anion $S^-$ in any reaction stage in which it is involved.

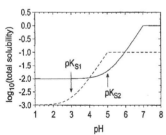

**Fig. 6.4** Manipulating relative solubilities.
For solid $S_1$,
    free acid solubility = 0.001M;
    anion solubility = 0.1M.
For solid $S_2$,
    free acid solubility = 0.01M;
    anion solubility = 1M.
$pK_a$ of acid $S_1$ = 3;
$pK_a$ of acid $S_2$ = 5

**Parallel ionisation processes: manipulating relative solubilities**

When two or more ionisable substances are involved in a reaction or workup process, this approach can be used to determine the relative amounts of each substance in solution. If two acidic solids having the same intrinsic solubility but different acidities are equilibrated together in the same aqueous solution with control of the pH, then the relative amounts of the two in solution will be a function of pH. The two curves of solubility vs pH can be plotted together to see at a glance the relative amounts of each in solution. Fig.6.4 shows the characteristics of one system. It is evident that by choice of appropriate pH the relative amounts in solution can be manipulated over a wide range.

## 6.3 Liquid–liquid partition of ionisable substrates

Ionisation processes may also be coupled to partitioning of a species between two solvents. For an acid SH partitioning between two liquid phases there are two equilibria to consider. The neutral form of the acid is partitioned between the organic solvent and the aqueous phase, and the acid is ionised in the aqueous phase to a degree which depends on the pH. It is assumed that the anionic form is insoluble in the organic solvent and that the pH is controlled by the addition of acid or alkali as necessary.

Quantitative treatment is again quite simple and requires the use of the relevant mass balance and equilibrium relationships. Partition of the free acid between water and the organic base is given by (6.9), where $K_D$ is the partition coefficient.

$$K_D = \frac{[SH_{org}]}{[SH_{aq}]},$$ 

(6.9)

Dissociation of the acid in the aqueous phase is given by (6.10).

$$K_A = \frac{[S_{aq}^-][H^+]}{[SH_{aq}]}$$ 

(6.10)

The mass balance must include both forms and locations of the acid species. Here it is assumed for simplicity that the phase volumes are equal (eqn. 6.11); if not, the appropriate weighting factor must be used.

$$[S_T] = [SH_{org}] + [SH_{aq}] + [S^-]$$ 

(6.11)

The concentrations are based on the volume of the specified phase, not on the total volume of the system

Solving these for the fraction of the total acid in the aqueous phase ($[SH_{aq(T)}] = [SH_{aq}] + [S^-]$) gives the relationship 6.12.

$$\frac{[SH_{aq(T)}]}{[S_T]} = \frac{1}{1 + \dfrac{K_D}{\left\{1 + \dfrac{K_A}{[H^+]}\right\}}}$$ 

(6.12)

Fig. 6.5 shows the characteristics of this system, where the extraction curves are plotted for acids with a partition coefficients of 1 and $10^3$. ($pK_D = 0$ and 3). Also shown is the ionisation curve vs pH for the acid itself in water in the absence of solvent.

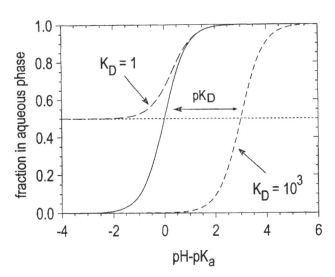

pK$_D$ is here defined as $-\log_{10}K_D$.

Note that the pH for half extraction is not the same as the pK$_a$. When K$_D$ is greater than about 10 the extraction curve is displaced to higher pH by approximately pK$_D$.

**Fig. 6.5** Plot of fraction of total acid in the aqueous phase vs pH for acids with a partition coefficients of 1 and $10^3$. The full line shows the fraction ionised in solution vs pH-pK$_a$.

**Manipulating relative distribution behaviour**

When two components of a mixture have different acidities, their distribution between two phases will show different dependencies on pH. This effect can be used to effect a separation

## 6.4 Ternary phase diagrams

In the systems considered so far the solvent properties of each phase have been treated as constant (with the exception of a change in pH). In three component (ternary) mixtures it is not uncommon for the interactions to be more complex, so that the properties cannot be simply described in two-component interactions. Triangular diagrams are a convenient way of representing the properties of these systems. The three pure components exist at the apices of the triangle. At the side opposite the apex the concentration of the apical component is zero. Lines parallel to this side have equal concentrations of the apical component. At any point on the diagram the fraction of any component present is represented by the distance from the side opposite the apex to this line. Fig.6.6 shows these properties.

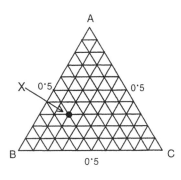

**Fig. 6.6** Properties of a triangular diagram. Point x has the composition A = 0.3, B = 0.5, C = 0.2.

A useful property of these diagrams is that, if more of one of the components is added to a ternary mixture, the composition moves along a line connecting the apex representing the pure component being added to the initial composition.

An example of the use of these diagrams came in the hazard analysis of a diazotisation process. One process for the manufacture of fluoroaromatic compounds involves the diazotisation of an amine in hydrofluoric acid by gradual addition of solid sodium nitrite (Fig. 6.7), prior to thermal decomposition of the derived diazonium salt (Fig. 4.1).

$$ArNH_2 + NaNO_2 + 2HF \xrightarrow{\text{HF/H}_2\text{O}} ArN_2^+F^- + NaF + 2H_2O$$

**Fig. 6.7** Diazotisation of an arylamine.

If diazotisation is attempted under conditions where the solution is saturated in sodium fluoride, then the sodium nitrite may not dissolve completely, due to coating by fluoride salts. This is potentially hazardous, because it can lead to build-up of unreacted material in the reaction mass. If, at a later stage of the process, the conditions are changed such that the coating can dissolve, then the accumulated sodium nitrite will dissolve rapidly, leading to a dangerous exotherm and vigorous boiling of the reaction mass.

The phase diagram for sodium fluoride/hydrogen fluoride/water (Fig. 6.8) is necessary to understand the nature of the problem.

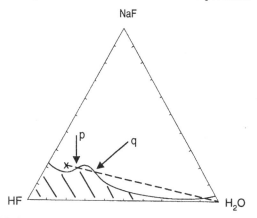

**Fig.6.8** Phase diagram for the system sodium fluoride/hydrogen fluoride/water at $0°C$.

The hatched area represents the desired single phase region. Outside this region a solid solvate will be present. So in the first instance the diagram enables a safe operating region to be defined. It is also helpful to be able to see what happens if a single component is added. For example, if to the composition represented by point **X** is added water, the composition will move towards the water apex, and will become homogeneous at point **P**. As more water is added the mixture will again become heterogeneous at point **q**. This type of analysis can thus be used to identifying safe operating conditions.

## Bibliography

Grant D.J.W. and Higuchi T. (1990). *Solubility behaviour of organic compounds*, Vol. XXI in 'Techniques in Chemistry,' Wiley-Interscience.
Hansch C. and Leo A. (1979). *Substituent constants for correlation analysis in chemistry and biology*, Wiley-Interscience.

# 7 Dispersion and mass transfer in multi - phase systems

Many reactions involve more than one phase, usually two, commonly three and even four or more. A typical example is presented in Chapter 2 Fig. 2.8. It is the catalytic hydrogenation of a nitrile to give a primary amine. If the nitrile is a solid with a solid catalyst, and the product amine is a solid, then, as hydrogen is obviously a gas and the reaction mixture is carried in a liquid solvent phase, there are five phases in all: three distinct solid phases, one gas and one liquid. In order for the reaction to proceed, mass must transfer between these phases, and it is easy to overlook the fact that considerable effort may be required to distribute them well enough and promote mass transfer to a sufficient extent that the reaction can proceed at a reasonable rate. This is important both in the laboratory and at large scale. For example 100kg of 5mm $Na_2CO_3$ agglomerates will dissolve in 1 $m^3$ water in a reactor as quickly as they are added if they are well dispersed, but will take 4 to 5 hours if allowed to settle on the vessel base. Similarly a gas bubbled through a liquid will not react at any meaningful rate unless well dispersed as small bubbles, and immiscible liquids will not react unless one is dispersed as fine droplets within the other. It is most common for fine chemical reactions to be carried out in a continuous liquid phase, which is assumed throughout this chapter, as vapour phase technology is not common for the generally high molecular weight, complex molecules considered here. They would require high temperatures to vapourise, outside the conditions suitable for much of the chemistry and where the materials tend to be thermally unstable .

## 7.1 Immiscible liquids - phase continuity

In a mixture of two immiscible liquids, it is possible for either of them to be the continuous phase with the other dispersed as drops within it. A mixture of oil and water can exist as water drops in oil or oil drops in water. It is not always recognised that dispersions of the same two liquids can behave differently depending upon which of them is the dispersed or droplet phase. *A dispersion of oil droplets in an aqueous phase can have very different properties to the same aqueous phase dispersed as drops in the oil.* There are many situations where the choice of phase continuity can seriously affect the process performance.

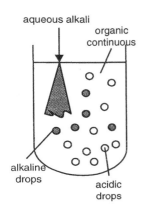

**Fig 7.1** The effect of phase continuity on reaction.

**7.1(a)** Aqueous phase dispersed. pH control depends on the ability of the dispersed drops to coalesce.

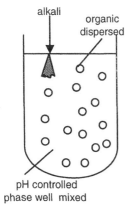

**7.1(b)** Organic phase dispersed. pH control simple in well mixed bulk phase.

The simplest example is pH control. Even the measurement of pH in a two phase system can be difficult. A pH probe measures the properties of the continuous phase. It will therefore measure the pH of an aqueous continuous phase with little problem, even when there may be discrete oil drops dispersed in it. If, however, the oil is continuous and the aqueous is dispersed as discrete drops, there is no longer a continuous aqueous phase to measure. Measurement of the pH of the aqueous phase would require the drops to coalesce and form a continuous phase around the probe, a most unlikely event. It is therefore almost impossible to measure the pH of a dispersed aqueous phase reliably.

Fig. 7.1 illustrates a practical problem of adding an alkali or acid to maintain pH at a controlled level during a reaction. For illustration consider a reaction where the reagents partition from the organic into the aqueous, where they react to generate protons. A number of examples are discussed in detail in chapter 8. Alkali is added to neutralise to maintain pH. In the first case (Fig.7a) the reactor is operated with the organic phase continuous and the aqueous dispersed. The alkali disperses as drops as it is added. The smallest and oldest droplets in the reactor will be acidic. In order for them to be neutralised they must first coalesce with the drops of fresh alkali dispersed in the reactor. The ability to control pH therefore depends on the coalescence rate. Slow coalescence will lead to a wide range of pH, from the high pH drops of fresh feed to the low pH un-neutralised drops.

In the second case (Fig. 7.1b), provided the mixing time is short the aqueous pH will be close to the desired value throughout the reactor. Here it is simply a matter of designing for good bulk mixing as described in chapter 5. In this example the aqueous continuous system will be the easier to design to reproduce laboratory performance.

Reactions between two immiscible liquids have additional complications. Not only can either phase be the dispersed or droplet phase, but in some circumstances the dispersed phase can switch to become continuous and vice-versa. This phenomenon is *phase inversion*. Fig. 7.2 illustrates the effect of phase inversion during a real industrial reaction. The reaction is fast and highly exothermic and temperature control is achieved by maintaining a high coolant flow to the reactor whilst controlling the reaction rate by restricting the rate of feed addition.

The reaction started with a low volume of low viscosity aqueous solution in the reactor and the mixed organic stream of medium viscosity was added steadily on temperature control (Fig 7.2a). Part way through the reaction it was observed that the inability to remove the heat of reaction even with the cooling fully on, caused the control system to reduce the feed rate dramatically, resulting in an extremely long and drawn out reaction. As well as very poor productivity, a very poor yield was obtained as the product went on to form by products. The cause was traced to the mixture changing unpredictably from a low viscosity aqueous continuous system to a high viscosity organic continuous emulsion, with appalling heat transfer properties (7.2b). This caused the difficulty in removing the heat of reaction and the consequent low feed rate. In order to overcome the problem, the reaction mixture had to be diluted with solvent. At full scale, the reaction could not be

**Fig. 7.2** Example of phase inversion during reaction. (Atherton J.H. (1993) Trans. I.Chem.E., Part A, 71)

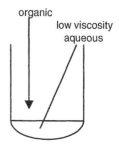

**Fig. 7.2(a)** Start of reaction Small volume of low viscosity aqueous solution (viscosity ($\mu$) =5*10$^{-3}$ Nsm$^{-2}$; density ($\rho$)=1144 kgm$^{-3}$)

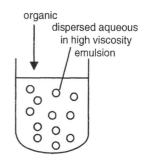

**Fig. 7.2(b)** End of reaction Mixed organic stream added. (Viscosity ($\mu$)=1Nsm$^{-2}$; density ($\rho$) =1208 kgm$^{-3}$ ). Final phase ratio 2 organic : 1 aqueous High viscosity emulsion of aqueous dispersed in organic.

operated reproducibly at the concentrations and phase ratios achieved in the laboratory.

## 7.2 Phase boundary diagrams and phase inversion

In order to predict and control which phase will be dispersed in a particular mixture, in any item of processing equipment, it is necessary to understand the effect of the controlling variables. The most important is volume fraction, which is often plotted to show regions of phase continuity in a phase boundary diagram, as illustrated in Fig. 7.3. The phases are distinguished by being the lighter or heavier phase, rather than aqueous or organic. There are three regions, one where the mixture will always be light phase continuous, one where it will be heavy continuous, and one where either phase can be continuous depending upon how the dispersion is formed. This is the *ambivalent region*. The parameters are *volume fraction* and for an agitated reactor, agitator speed or power.

The reaction example described above can be plotted on the phase diagram for illustration. Point A represents the starting point with the light phase in the reactor. As the heavier phase is added at constant agitator speed, its volume fraction increases until point B is reached. The upper boundary has no significance to a light phase dispersed system. At point B, so much dispersed phase is present that the drops are permanently in contact and the light phase can no longer form a continuous film. The drops coalesce and the heavy phase, which by this time has achieved the greater volume fraction, becomes continuous. This is the light continuous - heavy continuous (LC to HC) phase inversion boundary.

The mixture has a higher viscosity than the continuous phase, and in the example a high viscosity emulsion was produced when inversion occurred.

The dispersed phase is the phase present as drops in the mixture.

The continuous phase is the bulk phase in which the drops are dispersed.

$V_{fL}$ is the volume fraction of the light phase

$V_{fH}$ is the volume fraction of the heavy phase

Moving down the path A-B corresponds to adding heavy phase into the light phase and results in a light continuous to heavy continuous inversion at point B (in this case in the ambivalent region, the heavy phase is dispersed)

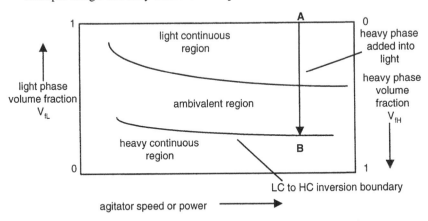

**Fig. 7.3** Phase boundary diagram (light continuous to heavy continuous inversion)

The analogous heavy to light inversion is shown in Fig. 7.4. Starting with the heavy phase in the reactor, point C, and adding light phase results in phase inversion at point D (HC to LC boundary). Here the lower boundary has no significance to a heavy continuous system.

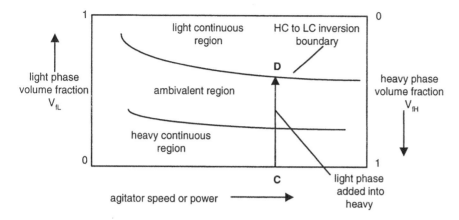

**Fig. 7.4** Phase boundary diagram (heavy continuous to light continuous inversion)

*In the upper region, the system can only ever be light continuous and in the lower region only heavy continuous. Between these boundaries either phase can be continuous depending upon the mode of operation.*

In these examples, the x axis of the phase boundary diagrams is agitator speed or power, but for the kind of agitated vessels most normally used for reactors, the phase boundaries will be relatively insensitive and other parameters largely determine continuity and phase stability.

Moving up the path C - D corresponds to adding light phase into heavy phase and results in a heavy continuous to light continuous inversion at point D (in this case in the ambivalent region the light phase is dispersed)

*Phase stability*

Many system properties can affect continuity at the phase boundaries or within the ambivalent region. Close to the boundaries, the system is most sensitive to *volume fraction* of the dispersed phase. Where there is no significant retardation of coalescence by surface active materials, the boundaries will be at a dispersed phase volume fraction around the maximum packing for spheres, around 60 to 65%. That is, even for non-stabilised systems *it is commonly possible to disperse up to 65% of the total volume in the remaining 35%.* In the laboratory it is common to operate with 50:50 mixtures, unfortunately this is often in the ambivalent region and therefore continuity depends upon the operation as much as the physical properties. Up to the inversion boundary, *adding a phase with the agitator running will cause it to be dispersed. Starting the agitator from rest will normally result in the phase surrounding it, normally the heavy phase, becoming continuous.*

Fig. 7.5 illustrates a common cause of operating problems in taking a process to full scale when the process demands a light continuous system to reproduce laboratory performance.

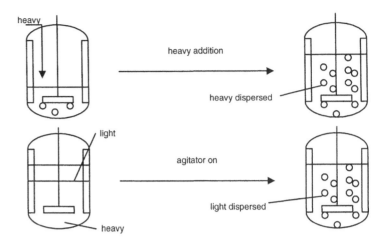

**Fig.7.5** The effect of operation on phase continuity

Adding the heavy phase to the light phase results in a light-continuous, heavy-dispersed system as required. If the agitator is stopped and the liquids are allowed to separate; or the liquids are added to the reactor with the agitator stopped, then when the agitator is started, the heavy phase will become continuous. Because the process requires a light continuous dispersion to achieve laboratory performance, there is a danger that heavy continuous will not give the required rate or selectivity.

A common problem in real industrial systems, which can be arise from generating the wrong dispersion is when one dispersion coalesces much more slowly than the other. It is caused by solutes or impurities retarding coalescence when they are in the continuous phase, but having no effect when on the dispersed side of the interface. A clean solvent continuous phase will commonly coalesce rapidly but the equivalent aqueous continuous system will coalesce more slowly. An extreme example is quoted in Table 7.1. It is the extraction of an organic acid from an aqueous mixture which has been produced by a biotransformation reaction. The aqueous contains a considerable amount of organic matter derived from the biocatalyst, which retards coalescence when it is present as the continuous phase. Conversely, the solvent phase is clean and therefore drops dispersed in it coalesce rapidly. In order to achieve a rapid clean separation after extraction, the mixture must be aqueous dispersed in solvent. Generating the wrong dispersion, solvent in aqueous, results in a vessel full of stable un-processable emulsion.

The coalescence properties of a dispersion are mainly dependant on the properties of the continuous phase. Surface active material in the continuous phase can severely retard coalescence.

**Table 7.1** Physical properties of an organic acid extraction, illustrating the effect of phase continuity on coalescence rate

| Parameter | Value |
|---|---|
| Aqueous density | 1060 kgm$^{-3}$ |
| Organic density | 800 kgm$^{-3}$ |
| Separation time (under gravity) aqueous continuous | 15 hrs |
| Separation time (under gravity) organic continuous | 150s |
| Volume fractions at organic continuous to aqueous continuous inversion | 0.6 (o) ; 0.4 (a) |

*Laboratory practice*

When developing a two liquid phase reaction or extraction in the laboratory any differences between the alternative dispersions should be identified and the phase boundaries established.

Operating close to the phase boundaries will always be less robust than away from the ambivalent region, as local changes in volume fractions can lead to unpredictable phase inversion. It is preferred to remain in the LC or HC regions, where the system cannot invert. The following rules of thumb should prove useful:

**Table 7.2** Phase stability rules of thumb

| General rules of thumb |
| --- |
| Non-stabilised dispersions will have ambivalent region boundaries around 60 – 65% by volume dispersed phase. |
| Increasing the agitator power destabilises a heavy continuous dispersion (in the ambivalent region near the inversion boundary). |
| Decreasing the agitator power destabilises a light continuous dispersion (in the ambivalent region near the inversion boundary). |
| At laboratory scale using materials of construction wetted by the dispersed phase can cause a dispersion (in the ambivalent region near the boundary ) to invert prematurely. |
| High viscosity liquids prefer to be dispersed. |
| Starting the agitator from rest will cause the phase around it to become continuous (at least initially). |
| The presence of air bubbles can cause an organic dispersed system to invert. |
| Details of the vessel and agitator design will have little effect on the phase boundaries but can effect the behaviour within the ambivalent region. |

## 7.3   Creating a dispersion - the just dispersed condition

During process development it should be established which dispersed phase gives best performance and the volume fraction to use for maximum rate, selectivity and ease of separation. It is then necessary to ensure that at full scale, the intended dispersed phase is distributed throughout the continuous phase without any separated layer. Fig. 7.6 shows the effect of increasing the agitator speed in a stirred vessel containing two separated phases. In Fig. 7.6, as the speed is increased, the light phase gradually disperses until there is none left as a separated layer. This is the minimum condition for design and is referred to as the "just dispersed speed" ($N_{jD}$). The effect of increased agitation on mass transfer above this limit is described later.

In a stirred vessel, the choice of agitator is important. In order to disperse two immiscible phases, it is essential that wall baffles and a turbine type agitator are used. They are described in chapter 5. Without baffles the liquids will never disperse well and poor performance will result. Fig. 7.7 shows the analogous situation for an unbaffled vessel and applies equally at any scale, i.e. *for reproducible performance even in the laboratory, a vessel, baffle and agitator system similar to Fig. 7.8 must be used.*

The commonly available laboratory round flasks with collapsible agitators are not suitable for two-phase reactions involving liquid-liquid and gas-liquid

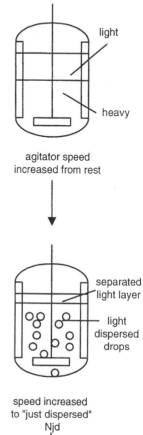

light

heavy

agitator speed
increased from rest

separated
light layer

light
dispersed
drops

speed increased
to "just dispersed"
Njd

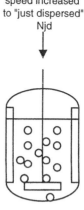

**Fig.7.6** Dispersing two liquids by increasing agitator speed in a baffled vessel

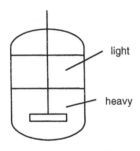

increased agitator speed
results in little dispersiion

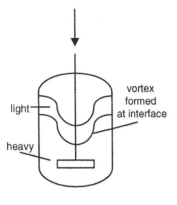

light

heavy

vortex
formed
at interface

**Fig 7.7** Poor and non-reproducible
dispersion without baffles

**Example of the speed and power
to disperse immiscible liquids:**
For a dispersion of 40% by volume
toluene in water. Toluene is light,
and dispersed. From equation 7.1:

$$\rho_c = 1000 kg m^{-3} ; \Delta\rho = 200 ;$$

$$\mu_c = 10^{-3} Nsm^{-2} \ V_{fd} = 0.4$$

$$\varepsilon = 0.12 W kg^{-1}$$

For a 10m³ reactor; power = 1.1kW
For a baffled turbine agitator with
Po=1, diameter 1.17m, from
chapter 5, speed

$$N_{jD} = 0.84 s^{-1} \ or \ 50 \ rpm$$

For a similar 1litre reactor:

$$N_{jD} = 7.47 s^{-1} \ or \ 450 \ rpm$$

For a similar dispersion of water in
toluene, from equation 7.2:

10m³  $\varepsilon = 1.76 W kg^{-1}; N_{jD} = 2 s^{-1}$

1 litre  $N_{jD} = 18.5 s^{-1}$

reactions. The interfacial area generated cannot be predicted theoretically and is not reproducible experimentally and such reactors cannot be used to generate reaction rate data which can be used for scale-up purposes. The preferred equipment is illustrated in Fig. 7.8. It can be made out of glass. The agitator diameter is half, and the baffle width is one tenth of the vessel diameter. Typical agitator speeds on this scale are in the region 5 – 15 s⁻¹.

$N_{jD}$ for full scale can be determined either by experiment or from physical properties using a design correlation. The experimental method requires a model reactor with a model agitator and internal fittings and liquids with the same density and viscosity as the real materials. Interfacial tension is important for drop size prediction but has little effect on the just dispersed speed. The preferred scale of operation is a vessel of 0.1 to 0.2 m diameter. From the observed speed at which the phases just disperse, calculate the agitator power, as described in chapter 5, dividing by the reaction mass gives the specific power input ($\varepsilon$). The full scale unit should be designed with the same specific power input. Empirical correlations are available to calculate the power from physical properties for light-in-heavy and heavy-in-light dispersions, which avoids the need for experiment:

light dispersed in heavy: $\varepsilon_{L-in-H} = 6.138 \times 10^{-7} \rho_c^{0.66} \Delta\rho^{0.96} \mu_c^{-0.4} V_{fd}^{0.24}$     (7.1)

heavy dispersed in light: $\varepsilon_{H-in-L} = 1.89 \times 10^{-3} \Delta\rho^{1.1} \mu_d^{0.21} V_{fd}^{0.36}$     (7.2)

(Lines P.C. (1990), I.Chem.E. Symp. Series 121, 167)

Throughout this chapter, $\rho$ is density; $\mu$ is viscosity; $\varepsilon$ is specific power input; the subscript $c$ refers to the continuous phase and $d$ to the dispersed phase. SI units are used throughout.

## 7.4    Dispersing solid particles

There are many examples of reactions involving solid reagents. They can either be heavier or lighter than the liquid phase. The more normal case is heavier, where the solid particles will sit on the base of the vessel, unavailable for reaction if insufficient agitation is provided

*Suspension of heavy particles*

For an agitated vessel, there is an agitator speed at which the solid particles are just kept moving from the reactor base. They are swept upwards, fall back and are re-suspended. This is the minimum design condition for the agitator and reactor. Above this there may be an increase in mass transfer rate, but it will be minor compared to suspending the solids away from the vessel base in the first place. An agitator generating axial flow pattern is preferred, directly down the centre of the reactor, sweeping the base and returning upwards at the walls, as shown in Fig. 5.5 (top). This is achieved by using an angled blade turbine impeller or preferably a hydrofoil as described in chapter 5. In order to obtain reproducible results, a similar flow pattern is required even in laboratory glassware

The required agitator speed ($N_{js}$) can be observed in a small scale model vessel using the real materials or model materials with matching physical properties, or calculated knowing the physical properties and the agitator and vessel design:

$$N_{js} = \frac{S v^{0.1} d_p{}^{0.2} \left(\dfrac{g\Delta\rho}{\rho_l}\right)^{0.45} X^{0.13}}{D^{0.85}} \quad (7.3)$$

$S$ is a constant which depends on the system geometry. A value of 5.8 can be used for the most commonly used angled turbine agitator; $v$ is kinematic viscosity ($10^{-6}$ m$^2$s$^{-1}$ for water); $d_p$ is particle diameter; $D$ is agitator diameter, commonly half or one third of the vessel diameter for angled turbines; $g$ is the acceleration due to gravity (9.81 m$^2$s$^{-1}$) and $X$ is (mass of solid / mass of liquid) *100.

At $N_{js}$ the particles will not be distributed evenly throughout the reaction mass, but will on average be more concentrated around the agitator and base regions, which is usually sufficient for good performance. Occasionally a dispersion will be required where the solids are evenly distributed throughout the liquid, but this will take considerably more power to achieve than the just dispersed condition.

*Preventing light particles from floating on the surface*

In some cases the solid particles are lighter than the liquid and therefore would float on the surface if not continually drawn in by the liquid motion. In the laboratory it is fairly easy to see and overcome the problem provided a model agitator and vessel are used. Generally for this type of problem a twin impeller system should be used with an upper impeller close to the liquid surface as illustrated in Fig.7.9.

## 7.5   Dispersing gas as bubbles

In order to cause a gas to react at any measurable rate with a liquid or slurry, it must be dispersed as fine bubbles, which can only be achieved with the correct design of vessel, gas disperser or sparger and agitator. The agitator should be a turbine-type, the vessels must have baffles and the gas is introduced underneath or just above the agitator. *Without the correct agitator design, without baffles or with the gas introduced elsewhere other than into the agitator, the gas will not disperse well.* A low reaction rate will result and the prolonged reaction may also give poor yield as competing reaction rates may become significant. In the laboratory, a vessel and agitator similar to Fig. 7.8 should be used. Addition of a gas is made *via* a small diameter (ca. 1mm) nozzle either just below or just above the agitator, towards the centre of the flask. Use of a frit is not necessary: the agitator will break up the gas into small bubbles.

**Example of the agitator speed required to suspend NaOH pellets**

To suspend 1te NaOH pellets in 10te water (i.e. X =10%)
S=5.8; D=1.17m
$d_p$ = 0.005m (5mm)
$v = 10^{-6}$ m$^2$s$^{-1}$; $\rho_l$ = 1000kgm$^{-3}$
$\Delta\rho$ = (130 – 1000)= 1130 kgm$^{-3}$;
g=9.81ms$^{-2}$

From equation 7.3: Njs = 1.76 s$^{-1}$ (105 rpm)

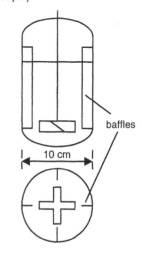

**Fig 7.8** Laboratory 1l vessel suitable for liquid-liquid or gas-liquid dispersion

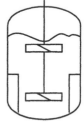

**Fig. 7.9** Twin impellers and short baffles for drawing down light solid particles

**Example of gas dispersion in a reactor:**

Could a chlorination requiring $0.167 m^3 s^{-1}$ gas be carried out successfully in a $10 m^3$ (2.4m diameter) reactor with a 1.17m diameter turbine agitator rotating at $1 s^{-1}$?

$Fr = Q/ND^3 = 0.119$; $Fl = 0.104$
For the reactor to work efficiently, equation (7.4) requires:
$Fl < 30*0.119*2^{-3.5} = 0.315$
$Fr = 0.119$, $< 0.315$; therefore the reactor will work.

However, equation (7.5) shows that to disperse bubbles throughout the whole reactor volume,
$Fl < 0.2(0.119)^{0.5}*2^{0.5} = 0.097$
i.e. gas will not be dispersed throughout the whole volume.

In reactions other than where there are fast competing by-product reactions, this would not be a problem.

Even with this correct equipment design, if the agitator speed is too low the gas will not disperse well enough. Fig. 7.10 shows the stages of dispersing a reacting gas in a liquid using a gas sparger and an agitator (A.W. Nienow, Proc. 5th Euro. Conf. Mixing, 1985). As the speed is increased the gas begins to disperse and the gas bubbles can be seen to fill the upper part of the reactor. The agitation rate at which this happens can be correlated with physical properties and agitator design as defined in equation 7.4. The dimensionless groups, the gas flow number (*Fl*) and Froude number (*Fr*) are used:

$$Fl \leq 30 Fr \left\{ \frac{T}{D} \right\}^{-3.5} \tag{7.4}$$

The gas flow number *Fl* is $Q/ND^3$ where *Q* is the gas rate, *N* is the agitator speed  and *D* the agitator diameter. *Fr* is the Froude number. It is proportional to the ratio of the forces acting on the bubbles due to the centrifugal action of the agitator and gravity; $N^2 D/g$. *It is essential for a large scale reactor to be designed in this way, if the reaction is to take place at a reasonable rate.*

As the agitation is increased even further the gas bubbles will be seen to fill the whole reactor. Again this can be correlated:

$$Fl \leq 0.2 Fr^{0.5} \left\{ \frac{T}{D} \right\}^{0.5} \tag{7.5}$$

In practise, although this condition is easy to achieve in the laboratory provided the correct agitator and vessel design and sparger arrangement are used, it is rather difficult at large scale and most commercial reactors will operate in the regime according to equation (7.4) (Fig. 7.10).

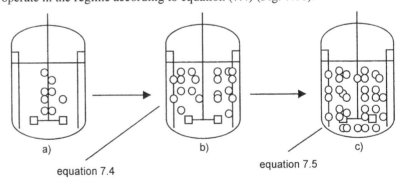

a)                          b)                          c)

equation 7.4                              equation 7.5

increasing agitator speed (fixed gas rate) ⟶

**Fig. 7.10** Increasing gas dispersion with increasing agitator speed or lower gas rate

### 7.6   Mass transfer

The dispersions described so far in this chapter all refer to the *minimum required for reasonable and reproducible reaction rate.* However it is possible to increase the reaction rate in systems where mass transport is the rate determining step by increasing the mass transfer rate above that achieved at this minimum condition. For solid particles the interfacial area is determined by the form of the particles and there is little benefit in increasing the agitation rate. For gases and immiscible liquids, however, increased agitation rate can generate smaller and smaller drops or bubbles with increasing interfacial area which will increase the mass transfer rate.

*Mass transfer in liquid dispersions.*
There are a number of ways of representing mass transfer rate across an interface. One of the simplest is the two film theory, by which the resistance to mass transfer is represented by two film resistances in series; a continuous phase film, and a dispersed phase film, as illustrated in Fig. 7.11. The rate of mass transport across an interface can be represented by :

$$J_i = K_c \Delta C^* \tag{7.6}$$

where $J_i$ is the molar flux (kmol s$^{-1}$m$^{-2}$ (interfacial area)) and $K_c$ is the mass transfer coefficient, in this case based on the continuous phase. It is similar to a kinetic rate coefficient. $\Delta C^*$ is the difference between the concentration of the transferring species in the continuous phase and the concentration which would be in equilibrium with the dispersed phase. Concentrations are molar units (kmol m$^{-3}$).

The rate equation (7.6) can be derived from a mass balance, assuming that in each phase: transport coefficient = rate coefficient * concentration driving force, where the driving force is the difference between the bulk concentration and the concentration at the interface (Fig 7.11).

dispersed phase: $J_i = k_d * (C_d^* - C_d)$ $\hspace{2cm}$ (7.7)

continuous phase: $J_i = k_c * (C_c - C_c^*)$ $\hspace{2cm}$ (7.8)

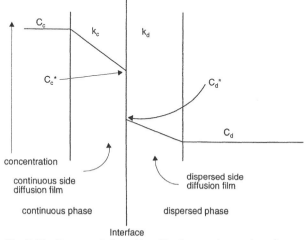

P is the partition coefficient of the transferring species: $P = \dfrac{C_c^*}{C_d^*}$

**Fig. 7.11**   Representation of two film theory of mass transfer

Equating (7.7) and (7.8), and eliminating the interfacial concentrations by assuming equilibrium at the interface and that a partition coefficient *(P)* can be defined for the transferring species such that:

$$P = \frac{C_c^*}{C_d^*} \tag{7.9}$$

Results in:

$$J_i = \frac{(C_c - PC_d)}{\frac{1}{k_c} + \frac{P}{k_d}} \tag{7.10}$$

$PC_d$ is the concentration which would be in equilibrium with the dispersed phase, thus $(C_c - PC_d)$ is an equilibrium driving force, $\Delta C^*$.

An overall mass transfer coefficient can be defined by summing resistances in series:

$$\frac{1}{K_c} = \frac{1}{k_c} + \frac{P}{k_d} \tag{7.11}$$

Hence equation (7.10) becomes the mass transfer equation (7.6).

$k_c$ and $k_d$ are the external (continuous phase) and internal (dispersed phase) film coefficients, which can be correlated to a reasonable approximation with physical properties and the fluid mechanics.

Where the transferring substance is present as a pure phase, there is no film resistance in that phase, and $1/k$ approaches 0. For gas mixtures or solutions the mass transport coefficients can be estimated from:

D is the diffusion coefficient of the transferring species; $d_{32}$ is the Sauter (area / volume) mean diameter of the bubble or drop.

$$k_d = \frac{2}{3}\pi^2 \frac{D}{d_{32}} \tag{7.12}$$

Sh is the dimensionless Sherwood number:
$Sh = k_c d_{32} / D$

$$k_c = Sh \frac{D}{d_{32}} \tag{7.13}$$

$Re_p$ is the dimensionless Reynolds number:
$Re_p = d u \rho_l / \mu$

Sc is the dimensionless Schmidt number:
$Sc = \nu / D$

All units are SI.

The dispersed phase film coefficient ($k_d$) is derived assuming diffusion through a stagnant drop, the continuous film coefficient ($k_c$ is based on the flow properties of the fluid around the drop (Rowe (1965)).

Rowe P.N. *et al.*, (1965), Trans. I.Chem.E. ,43, T14

$$Sh = 2 + 0.72 \, Re_p^{0.5} \, Sc^{\frac{1}{3}} \tag{7.14}$$

For most applications, the mass transfer coefficient will be used to predict the total flux between phases ($J$, kmol s$^{-1}$), or the rate of change of concentration in each phase:

$$J = A * J_i = K_c a V_c \Delta C^* \tag{7.15}$$

$$\frac{dC_c}{dt} = \frac{J}{V_c} = K_c a \Delta C^* \qquad (7.16)$$

Where $a$ is the interfacial area/continuous phase volume:

$$a = \frac{6V_{fd}}{d_{32}(1-V_{fd})} \qquad (7.17)$$

For a dispersion of 100 µm drops at a volume fraction of 0.4,
$a = 4*10^4 \ m^{-1}$

For a typical dispersion of 1mm bubbles at a volume fraction of 0.2,
$a = 2.5*10^3 \ m^{-1}$

*Mass transfer in gas - liquid reactions*

Gas - liquid mass transfer can be modelled in an analogous way to liquid dispersions. In this case the continuous phase is liquid, with gas dispersed. The partition coefficient is replaced by Henry's constant for the mixture ($He$) (subscript $l$ refers to the continuous liquid phase and $g$ to the dispersed gas):

$$\frac{1}{K_l} = \frac{1}{Hek_g} + \frac{1}{k_l} \qquad (7.18)$$

From Henry's law; the Henry constant for a mixture ($He$) is the ratio of the concentration in the gas phase to the concentration in the liquid and strictly applies to dilute solutions.

For many systems, $1/Hek_g \ll 1/k_l$ and $K_l \approx k_l$. It is difficult experimentally to separate $k_l$ and $a$, so they are usually correlated together. Most industrial systems can be approximated by the correlation presented by Middleton (1985) for "non - coalescing" systems, relating the mass transfer to the energy dissipation rate ($\varepsilon$) and the superficial velocity of the gas ($V_s$):

$$K_l a = 2.3\varepsilon^{0.7}V_s^{0.6} \qquad (7.19)$$

Middleton J.C., (1985), in "Mixing in the Process Industries", Eds. Harnby et al, Buttrerworth, London

From this correlation it can be seen that *small scale experiment can model full scale operation by providing the same power/mass ($\varepsilon$).*

## 7.7 Drop size and interfacial area

It is difficult to correlate drop and bubble size, hence interfacial area, with physical properties, because of the unpredictable effects of very small quantities of interfacially active material. However, there are correlations available which have been developed from experiments on model systems, and which provide a reasonable basis for designing small scale experiments to predict full scale performance.

*Drop and bubble sizes*

The analysis used to derive drop sizes in liquids can be applied to any flow regime, but the most common is turbulent flow (see chapter 5). The maximum size of a drop which can survive without being broken apart by the fluid energy has been correlated by Davies (1987) for a wide range of equipment:

Davies J.T., Chem. Eng. Sci., 42, 7, 1671 – 1676

$$d_{max} \propto \varepsilon_{max}^{-0.4}\rho^{-0.6}\sigma^{0.6} * (\mu_d correction) \qquad (7.16)$$

The viscosity correction term accounts for high viscosity dispersed phases and can be omitted for low viscosity liquids.

Therefore *the mean drop size and interfacial area can be related to the average (power input/mass)*, which forms the basis for predicting full scale performance from lab scale experiments in a model reactor.

Mean bubble sizes can be correlated in a very similar way.

# 8 Mass-transfer and reaction in two-phase systems

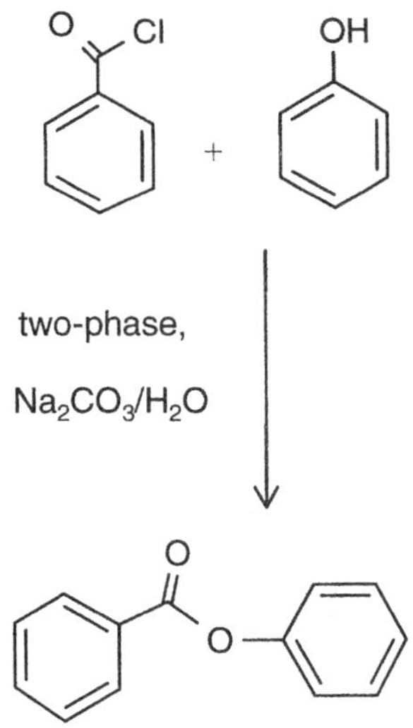

Two-phase reaction systems are sometimes used because they provide an intrinsically better selectivity than is achievable under homogeneous conditions, but their use may be also enforced by the nature of the reactants, if they are mutually insoluble. For large scale use, water immiscible solvents are often preferred because of their ease of recovery.

Some improbable looking reactions can be carried out successfully in two-phase systems. The well known Schotten-Baumann reaction - acylation of an amine or a phenol in a two-phase aqueous/organic system (Fig. 8.1) - can be conducted in high yield despite the fact that the half-life of benzoyl chloride in water is less than a second. Phosphoryl and thiophosphoryl chlorides can be similarly reacted with phenols in high yield to give the corresponding phosphites and thiophosphites. Acylated cyanohydrins can be prepared in a 'one pot' process by reacting aldehydes and cyanide ion with an acylating agent. These and more examples are listed in Table 8.1

**Fig. 8.1** The Schotten-Baumann reaction: benzoylation of phenol in a two-phase organic/aqueous system

**Table 8.1.** Some synthetically useful two-phase reactions

| Reaction | Product | Area of use |
|---|---|---|
| Catalytic chiral hydrogenation using homogeneous catalysts | Single enantiomer hydrocarbons and acylated amino-acids. | Intermediates for pharmaceuticals and agrochemicals. |
| Acid chloride with phenolate salts | Esters | General organic synthesis |
| Acylating agent + aldehyde + CN⁻ | Acyl cyanohydrins | Intermediates for pharmaceuticals and agrochemicals. |
| Dehydrochlorination of 3,4-dichlorobutene with 20% NaOH | 2-chlorobutadiene | Polymer precursor |
| Phenolate salts with thiophosphoryl chloride | Arylthiophosphates | Intermediates for agrochemicals |
| Alkyl or aryl halide + inorganic anion e.g. $CN^-$, $N_3^-$, or $F^-$ | Alkyl or aryl cyanide, azide or fluoride | Intermediates for pharmaceuticals and agrochemicals. |
| Benzyl cyanide with alkyl bromides/50% NaOH | $\alpha$-alkylbenzyl cyanides | Intermediates for agrochemicals |
| Addition of dichlorocarbene to olefins using chloroform and 50% NaOH. | Dichlorocyclopropanes | Intermediates for agrochemicals |

It is the purpose of this chapter to give an introduction to the interactions between physical and organic chemistry which are so important in determining the overall performance of two-phase reactions.

## 8.1 Reaction mechanisms

For practical purposes there are two mechanisms by which reaction may occur. Uncatalysed or extractive reactions occur following simple partition of the reactive component from a source phase to one where it is reactive. Other processes, in which a catalyst is required to facilitate interphase transport of the reactive species, are known as phase-transfer catalysed reactions. True interfacial reactions, i.e. reactions occurring within a nanometre or so of the interface, are rare. The extraction of $Cu^{2+}$ from aqueous solution using water insoluble ligands (Fig. 8.2) is an example of an interfacial process. Hydrolysis of the very reactive triphenylmethylchloride ($Ph_3C\text{-}Cl$), as a solid or as a solution in an organic solvent, is also interfacial.

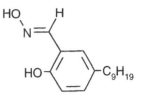

**Fig. 8.2** An oxime derived ligand used to extract $Cu^{2+}$ from water.

## 8.2 Extractive reaction

This is the most common mechanism in two-phase systems. It refers to a reaction which occurs following simple diffusion of a reactant A across a phase boundary from an organic phase in which it is stable to an aqueous phase where it reacts (Fig. 8.3). Depending on the rate of reaction, the chemical reaction may occur in the bulk aqueous phase or, if reaction is very fast, in the thin diffusion film adjacent to the interface

The locus of the reaction depends on the relative rates of reaction and diffusion of the reactive component in the phase where reaction occurs. Even in a well stirred and dispersed two-phase system there is a thin unstirred layer (the diffusion film) of roughly 50–100μm in thickness on either side of a liquid/liquid interface (Chapter 7). The transport time across this film is of the order of 5–10s. Reactions which occur faster than this will occur within the diffusion film and the reactant will never reach the bulk aqueous phase.

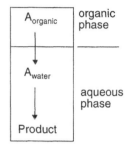

**Fig. 8.3** Extractive reaction

Mass transfer of gases is much quicker, and these approximate timescales should not be used for gas/liquid reactions.

### Reaction slow relative to transport across diffusion film

Reactions which occur on a timescale much longer than the transport time will occur in the bulk aqueous phase. Before this can happen the reactant must cross the organic/water interface, so the overall reaction rate is dependent on two processes which occur in series. Fig. 8.4 shows a model of this situation. A neat reactive organic liquid **A** (for example, an ester) is in contact with an aqueous phase containing a reagent **B** (for example, hydroxide) with which it reacts in the aqueous phase. The driving force for transport of **A** from the organic to the aqueous phase is the difference between the equilibrium and actual concentration of **A** in the aqueous phase.

The rate 'r' of the chemical reaction in the bulk aqueous phase is given by

$$r = k_1 C_{aq} \qquad \text{mol cm}^{-3}\,\text{s}^{-1} \qquad (8.1)$$

and the transport rate by

$$r = k_L a(C_{sat} - C_{aq}) \qquad \text{mol cm}^{-3}\,\text{s}^{-1} \qquad (8.2)$$

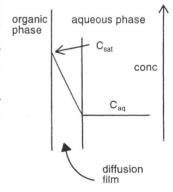

**Fig. 8.4** Model used to derive eqn. 8.3

The system is assumed to be at equilibrium across the interface. So when the organic phase is a neat liquid the concentration in the aqueous phase at the interface is assumed to be equal to the solubility, $C_{sat}$.

where $k_1$ is the first order rate constant in $s^{-1}$, $k_L$ is the mass transfer coefficient in $cm\,s^{-1}$, 'a' is the interfacial area per unit volume of the aqueous phase, $cm^{-1}$, $C_{aq}$ is the concentration of **A** in the aqueous phase and $C_{sat}$ is the equilibrium solubility of **A** in the aqueous phase. Since at any time these two rates must be equal, (8.1) and (8.2) can be solved simultaneously to give the overall rate in terms of known parameters:

$$\frac{C_{sat}}{r} = \frac{1}{k_1} + \frac{1}{k_L a} \tag{8.3}$$

The overall reaction rate does not increase indefinitely as the interfacial area increases, but reaches a plateau. There are two extreme cases depending on the relative values of $k_1$ and $k_L a$.

When the chemical reaction rate is much faster than mass transfer, so that $k_1 \gg k_L a$, then eqn. 8.3 reduces to (8.4)

$$r = k_L a C_{sat} \tag{8.4}$$

This situation is referred to as *mass transfer controlled.*

For the case where $k_L a \gg k_1$, eqn. 8.3 now reduces to (8.5)

$$r = k_1 C_{sat} \tag{8.5}$$

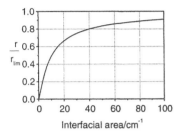

**Fig. 8.5** Reaction rate **r** relative to the chemically limited rate $r_{lim}$ vs interfacial area for eqn. 8.3. $k_1 = 10^{-2}\,s^{-1}$; $k_L = 10^{-3}\,cm\,s^{-1}$.

and the overall reaction rate is said to be *chemically rate limited.*

Fig. 8.5 shows the response of the overall reaction rate to interfacial area for eqn. 8.3 with some example values of the constants.

When the organic reactant is present in a solvent the treatment is slightly more complicated and the diffusional resistances on each side of the interface must be taken into account. This is discussed in Chapter 7.6 and also later in this Chapter.

Note that in eqn. 8.7 the units are those of a flux rate. Great care is needed when calculating rates in two-phase systems to ensure that the units are consistent. A check on dimensionality should always be carried out.

### Reaction fast relative to transport across diffusion film

When reaction is fast enough to occur within the diffusion film then diffusion and chemical reaction are coupled differently. The reactant concentration profile in the film is given by eqn. 8.6.

$$\frac{C_x}{C_{sat}} = \exp\left(-x\sqrt{\frac{k_1}{D}}\right) \tag{8.6}$$

and the reaction rate by eqn. 8.7

$$j = C_{sat}\sqrt{Dk_1} \quad mol\,cm^{-2}\,s^{-1} \tag{8.7}$$

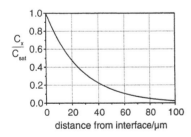

**Fig. 8.6** Concentration/distance profile for the reactive diffusion of benzoyl chloride into water (The calculation is approximate since the required diffusion coefficient in water has to be estimated)

where $C_x$ is the concentration at distance x from the organic/water interface, D is the diffusion coefficient of the reactive component in the reacting phase, and j is the reaction flux. Fig. 8.6 shows the concentration/distance profile for the reactive diffusion of benzoyl chloride into water. Reaction is complete in a thin surface layer adjacent to the interface. Very little benzoyl chloride reaches the bulk aqueous phase. For this type of reaction the reaction rate is thus directly proportional to interfacial area.

It is well known that the Schotten-Baumann reaction must be carried out with good agitation or vigorous shaking. This is because reaction between the acid chloride and the phenolate anion occurs in the aqueous phase diffusion film. Good agitation is necessary to ensure that the pH in the interfacial zone is maintained high enough to ensure that the reactive phenolate ion is present in preference to the much less reactive free phenol, so as to minimise competitive solvolysis of the acid chloride by water. The principles discussed in Chapter 4.4 are highly relevant here.

Because the reaction rates are proportional to the solubility of the dissolving component in the aqueous phase, unusual effects can occur when the aqueous phase reagent concentration is increased. In the hydrolysis of formate esters with sodium hydroxide, the hydrolysis rate *decreases* as the hydroxide concentration increases (Fig. 8.7). This is mainly because the salting out effect as the hydroxide concentration is increased, decreases $[C_{sat}]$ to a greater extent than the pseudo first order rate constant is increased. This effect is important in reactions where highly concentrated (50%) sodium hydroxide solution is used as a reagent (section 8.3). The salting out effect (Chapter 6) is so great that competitive hydrolysis of alkyl and even highly reactive acyl halides is not a problem.

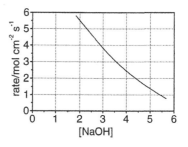

**Fig. 8.7** Response of hydrolysis rate of n-hexyl formate to [OH⁻]. Nanda A.K. and Sharma M.M. (1966). Chem. Eng. Sci., 21, 707

### Comparison of 'fast' and 'slow' reactions

Table 8.2 summarises and compares the behaviour of the two types of reaction. Note that for pseudo first order reactions where B is the reagent in excess, the rate is proportional to $\sqrt{[B]}$ rather than [B].

**Table 8.2** Response of overall reaction rate of 'in-film' and bulk phase reactions to changes in applied parameters

| Parameter | Effect on reaction rate | |
|---|---|---|
| | 'Fast' reaction | 'Slow' reaction |
| Interfacial area, 'a'. | Directly proportional | Asymptotic increase |
| Reactant concentration at interface | Directly proportional | Directly proportional |
| Chemical reaction rate constant, $k_1$ | Increases as $\sqrt{k_1}$, or as $\sqrt{k_2[B]}$ for pseudo first order reaction where B is in excess over A. | Asymptotic increase |

Two factors conspire to reduce reaction rates in an extractive reaction relative to homogeneous systems. First, the relatively poor solubility of many organic substrates in water means that the available concentration is low. Reaction rates are reduced pro-rata to the solubility (or partition coefficient), so a substrate ten times less soluble than another will react ten times more slowly. Second, the interfacial area will influence the overall rate, in a way which depends on the reaction regime.

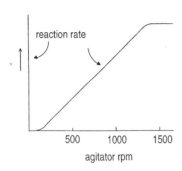

**Fig. 8.8** Typical response of reaction rate to agitation speed for a fast 'in-film' reaction, for a laboratory reactor of the type shown in Fig. 7.8.

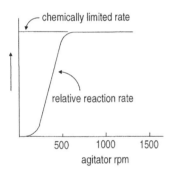

**Fig. 8.9** Typical response of reaction rate to agitation speed for a bulk phase reaction

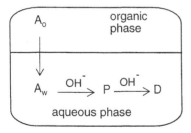

**Fig. 8.10** Consecutive processes in an extractive reaction

### Effect of agitation on reaction rate

In chapter 7 the mixing of two-phase liquid/liquid and gas/liquid mixtures was described. These principles can be usefully applied to reacting systems. The rate of a reaction in the 'fast' regime will increase in proportion to the interfacial area. A typical rate profile is shown in Fig. 8.8. The rate changes little until the upper phase is drawn into the region of the agitator blades, whereupon dispersion of the upper phase into the lower occurs. At this point a sudden increase in rate is apparent. Further increase in agitator speed increases the reaction rate until the point at which droplet coalescence or cavitation of the agitator by dispersed gas prevents any further increase in area.

The 'slow' reaction behaves differently. Again, little happens to the reaction rate until 'pickup' occurs, after which the rate increases until the reactive phase contains the equilibrium amount of reactant (Fig, 8.9). At first glance Figs. 8.8 and 8.9 appear similar; in reality, the factors which control the shape of the curves after the 'pickup' point are completely different.

In practice, agitation conditions can have a major effect on the rate and selectivity of multiphase reactions. The interfacial area per unit volume in a stirred vessel can vary over an enormous range, from 0.1 cm$^{-1}$ in a vessel containing an unstirred liquid of 10 cm depth., to several hundred cm$^{-1}$ in a well stirred laboratory vessel. Use of properly designed equipment (Chapter 7) is necessary to obtain meaningful results. At worst, use of inappropriate equipment may lead the experimentalist to conclude that a given reaction is not feasible, whereas the problem could simply be that the mass transfer rate is insufficient to give a measurable reaction rate under the conditions employed.

Solid/liquid mixtures behave differently from two-fluid systems in that the interfacial area is independent of the agitation rate. For small particles (diameter < 10 μm), once the solid is fully suspended in the reactor further increase in the agitation rate has little effect on the reaction rate. This is because mechanical agitation has little effect on the fluid flow close to the surface, and transport from the surface to the bulk is diffusion limited. All that agitation does is to provide a constant composition for the bulk phase in which the solid is suspended. If the agitation is so poor that the solid is not fully lifted from the vessel base, the liquid phase in the region of the solid particles becomes depleted of reactant and so reaction slows down.

### How heterogeneity can influence selectivity

The principle is simple: changes in reactant/reagent supply rates caused by variations in interphase transport rates can influence relative reaction rates. When the product-forming reaction is favoured by an increase in mass transfer rate, the selectivity is increased, and vice-versa. A simple model can be used to illustrate the principle for a 'slow' reaction. Fig. 8.10 shows the case of an extractive reaction where the product **P** is water soluble and further reacts to **D**. This could be exemplified by the hydrolysis of an ester bearing other sensitive functionality. The slower the transfer and reaction of component **A**, the longer product **P** dwells in the aqueous phase in the presence of hydroxide ion, and consequently the more by-product **D** which is formed.

Fig. 8.11 shows the model used to calculate the selectivity.

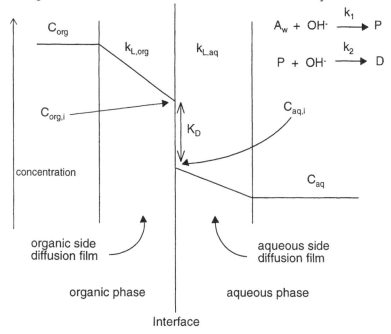

$$A_w + OH^- \xrightarrow{k_1} P$$

$$P + OH^- \xrightarrow{k_2} D$$

$k_{L,org}$ = mass transfer coefficient in the organic phase
$k_{L,aq}$ = mass transfer coefficient in the aqueous phase
$K_D$ = the partition coefficient for component A between the organic solvent and water (section 6.1), so at the interface,

$$\frac{[C_{org,i}]}{[C_{aq,i}]} = K_D$$

**Fig. 8.11** Model for quantitative treatment of competing processes in an extractive reaction

The pseudo steady state assumption is applied to the transfer of A across the interface and reaction in the aqueous phase. For equal phase volumes:

$$-\frac{dC_{org}}{dt} = ak_{L,org}(C_{org} - C_{org,i}) = ak_{L,aq}(C_{aq,i} - C_{aq}) = k_1 C_{aq}[OH^-] \quad (8.8)$$

where 'a' is the interfacial area per unit volume of the aqueous phase. Elimination of the interfacial concentrations using the partition coefficient gives eqn 8.9; eqns 8.10, 8.11 and 8.12 follow.

The derivation follows the same procedure as in 7.6. Both reactions are taken to be second order, and equal volumes of organic and aqueous phases are assumed.
Again, the concentration of reactant A is denoted by C

$$\frac{dC_{org}}{dt} = \frac{-C_{org}}{\dfrac{1}{a}\left\{\dfrac{1}{k_{L,org}} + \dfrac{K_D}{k_{L,aq}}\right\} + \dfrac{K_D}{k_1[OH^-]}} \quad (8.9)$$

$$\frac{dC_{aq}}{dt} = -\frac{dC_{org}}{dt} - k_1 C_{aq}[OH^-] \quad (8.10)$$

$$\frac{dP}{dt} = k_1 C_{aq}[OH^-] - k_2 P[OH^-] \quad (8.11)$$

$$\frac{dD}{dt} = k_2 P[OH^-] \qquad (8.12)$$

$$\frac{d[OH^-]}{dt} = -k_1 C_{aq}[OH^-] - k_2 P[OH^-] \qquad (8.13)$$

Table 8.3 summarises the characteristics of the system using some exemplary constants and a molar ratio of ester to hydroxide of 1:1, and Fig. 8.12 shows how the yield varies with interfacial area.

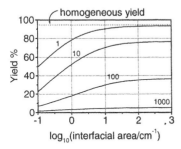

**Fig. 8.12** Effect of partition coefficient $K_D$ (shown) and interfacial area on yield for the example given in Table 8.3.

**Table 8.3** Limiting yields at high interfacial area and corresponding reaction times for a two-phase reaction with a competing consecutive process.

| $K_D$ | Yield % | $t_{99\%}$ (hr) |
|---|---|---|
| Homogeneous | 94.7 | 1.25 |
| 1 | 93.5 | 1.4 |
| 10 | 76.6 | 4.4 |
| 100 | 36.4 | 17 |
| 1000 | 5.2 | 110 |

$k_1 = 10^{-2} \ M^{-1}s^{-1}$; $k_2 = 10^{-4} M^{-1} s^{-1}$; $k_{L,org}$ and $k_{L,aq}$ have both been taken as $10^{-3}$ cm s$^{-1}$. Yield of P is that when all the hydroxide has been consumed.

Note that

- The rate and selectivities increase with interfacial area.
- Even at high interfacial areas, such that mass transfer is not rate limiting, the rate and selectivity are reduced relative to the homogeneous case in an amount dependent on the distribution coefficient, and
- When $K_D$ is large, unrealistically long reaction times are needed in the two phase system.

A corollary is that selective removal of a more soluble species from a mixture of two components *with the same chemical reactivity* can be achieved using the distribution coefficient as the discriminating parameter.

In the above example use of a two phase system is non-ideal. The optimum performance will be achieved by the use of a solvent to render the reaction homogeneous.

**Table 8.4** Solubility of hydrogen in common solvents, 20°C. 1 atm.

| Solvent | Solubility M x $10^4$ |
|---|---|
| water | 7.6 |
| ethanol | 36 |
| acetone | 29 |
| diethyl ether | 54 |
| benzene | 29 |

**A practical example: chiral hydrogenation.**

In the hydrogenation in methanol solvent of the prochiral olefin methyl (Z)-α-acetamidocinnamate using a chiral rhodium catalyst, the predominant product is the S-enantiomer (Fig. 8.13), but some R-enantiomer is formed in an amount which depends both on the hydrogen overpressure *and on* the agitation conditions in the reactor.

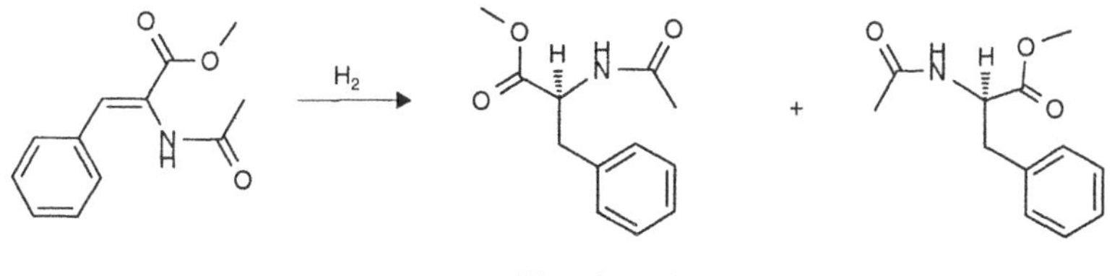

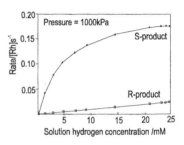

(S) major product        (R) minor product

**Fig. 8.13** Hydrogenation of methyl-(Z)-α-acetamidocinnamate.

This is because the enantioselectivity (determined by the relative rates of formation of the S- and R-enantiomers) depends on the hydrogen concentration *in solution* (Fig. 8.14), and this is not necessarily directly proportional to the hydrogen pressure. The hydrogen concentration depends on the relative rates of supply of hydrogen (by mass transfer) and consumption (by reaction). Hydrogen has only a low solubility in water and organic solvents (Table 8.4), and mass transfer is frequently rate limiting in reductions using hydrogen gas. Eqn. 8.3 is relevant here. For given reactant and catalyst concentrations there are two ways of achieving a given hydrogen concentration in solution: either by modifying the mass transfer rate at constant hydrogen overpressure (Fig. 8.15), or by varying the hydrogen overpressure with the mass-transfer conditions constant. In this example the enantioselectivity is *decreased* as the hydrogen concentration is *increased*.

The opposite effect is also seen. Hydrogenation of geraniol to citronellol shows *increased* enantioselectivity as the solution concentration of hydrogen increases (Fig. 8.16).

**Fig. 8.14** Rates of formation of S- and R-enantiomers (Fig. 8.13) *vs* solution hydrogen concentration.
Data from Landis C.R. and Halpern J. (1987). J. Amer. Chem. Soc., 109, 1746.

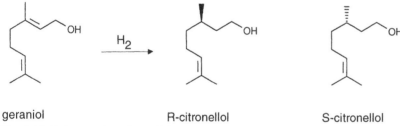

geraniol              R-citronellol              S-citronellol

**Fig. 8.16** Hydrogenation of geraniol

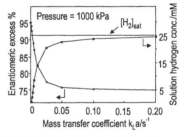

**Fig. 8.15** Enantioselectivity vs $k_La$ and solution hydrogen concentration for the hydrogenation of methyl-(Z)-α-acetamidocinnamate.

Figs. 8.14 and 8.15 redrawn with permission from: Sun Y., Landau R.N., Wang J., LeBlond C. and Blackmond D.G., J. Amer. Chem. Soc. (1996). 118, 1348.

*Thus in order to obtain meaningful and reproducible results in this type of catalytic enantioselective hydrogenation it is necessary to control the solution concentration of hydrogen.*

An appreciation of the fundamentals of heterogeneous reaction systems is essential in order to select equipment properly and design experimental work in this area.

## 8.3 Catalysed two-phase reactions - phase transfer catalysis

The general principles which have been described for extractive reaction apply also to catalysed reactions, but the systems are more complex, because of the additional dimension which is the nature and effect of the catalyst.

The original impetus behind phase transfer catalysis came from the ability to overcome the rate restrictions imposed by the mutual insolubility of inorganic reagents and organic reactants, but the scope has developed greatly since then. It is worth making the distinction between the original concept of phase transfer catalysis, as defined by Starks (Fig 8.17), which involves simple ion transport, and reactions which involve deprotonation of a weak base, usually a carbon-acid, followed by reaction of the derived carbanion (Fig 8.20). The latter are more common, are mechanistically more complex, and are more difficult to optimise. Some cations commonly used as phase-transfer catalysts are shown in Table 6.2.

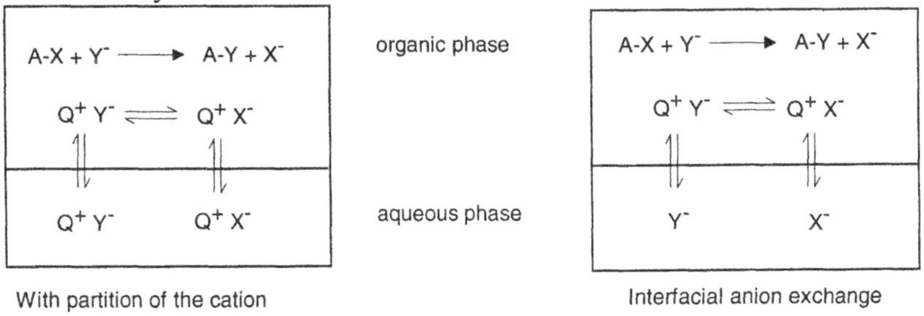

Fig. 8.17 Mechanism of simple phase-transfer catalysis (Starks C.M. (1971). J. Amer. Chem. Soc., 95, 3613).

## Simple phase-transfer catalysis

Two examples which exemplify the utility and unusual features of phase transfer catalysis are shown in Figs. 8.18 and 8.19.

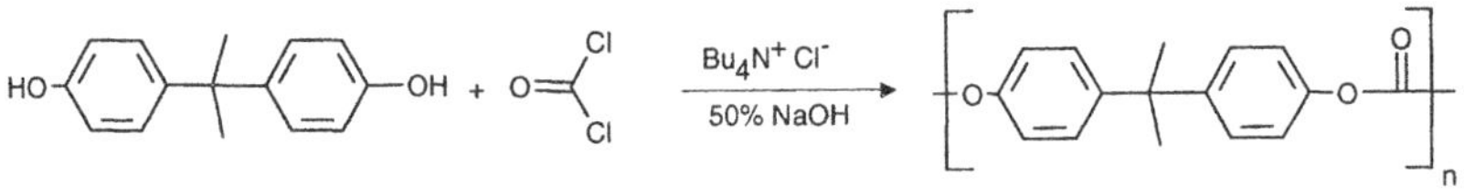

Fig. 8.18 Catalysed reaction of phosgene with a bis-phenol. (US patent 5,391,692 (1995). General Electric Co.)

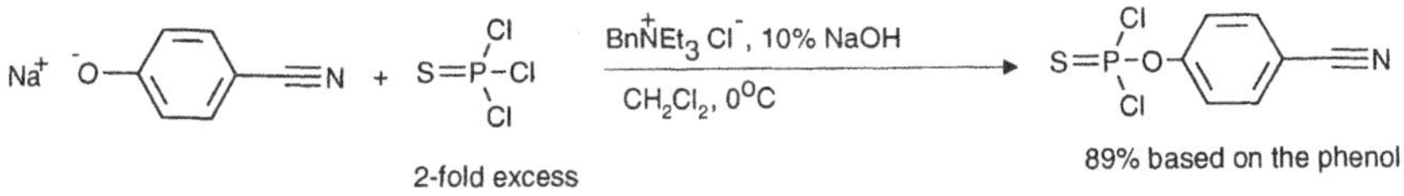

Fig.8.19 Catalysed reaction of thiophosphoryl chloride with phenols. (Shipov A.E., Genkina G.K, Mastryukova T.A. and Kabachnik M.I. (1993). Russ. J. Gen. Chem., 63, 1196

In both cases the remarkable feature is that readily hydrolysable acylating agents give a satisfactory reaction in the presence of water. The reason is that the acylating agents are hardly partitioned into the water phase under the reaction conditions, so that all the reaction occurs in the organic phase.

Again, reactions may be chemically or transport rate limited under sensible laboratory agitation conditions. A judgement can be made based on the nature of the chemistry. Reactions of alkyl halides with nucleophiles such as cyanide, carboxylates, azide are relatively slow. Acylations of hetero-anions such as sulphides, phenoxides and amide anion will, in contrast, be fast.

# PTC and carbanion chemistry

It is helpful to consider this class of reaction separately from PTC, since they are mechanistically distinct. Typically 50% aqueous sodium or potassium hydroxide is used to deprotonate a carbon acid, and the resulting carbanion is transferred to the organic phase as an ion pair. When a small non-lipophilic cation is used e.g. benzyltrimethylammonium, hydroxide cannot be transported into the organic phase and deprotonation occurs at the interface. Fig. 8.20 shows an example. With weakly acidic carbon acids the deprotonation step can become rate limiting, and the overall reaction rates are thus very sensitive to interfacial area. With larger more lipophilic cations hydroxide itself may be transported.

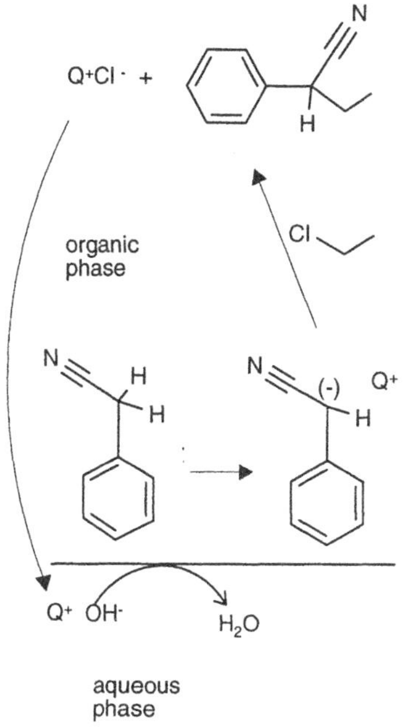

## Catalyst choice

### Solubility and partitioning of the catalyst

The catalyst must transport the desired anion from the aqueous to the organic phase. Large lipophilic cations e.g. methyltrioctylammonium salts are virtually insoluble in water and ion exchange occurs at the interface. Tetrabutylammonium and tributylmethylammonium salts are usually distributed between both aqueous and organic phases. Both classes can be used as catalysts. Tetramethylammonium salts are usually too water soluble to be useful in liquid-liquid systems but can be useful in solid-liquid reactions (section 8.4). Crown ethers form stable lipophilic complexes with alkali metal cations, and will transport anions into an organic phase.

### Catalyst stability

Strongly basic anions (hydroxide, fluoride), will decompose simple alkylammonium ions via the β-elimination. Several solutions to this problem have been proposed. Tetramethylammonium salts can be used, since the β-elimination is not now possible (but nucleophilic displacement of a methyl group can still occur, albeit under more stringent conditions). Phosphonium salts are more stable than the corresponding ammonium salts, but are too expensive for many applications.

**Fig. 8.20** Prior deprotonation in phase-transfer catalysis

### Catalyst separation from product

In many cases it will be necessary to remove the catalyst from the product in order to meet quality specifications or to avoid interference with subsequent reaction stages. This requirement can limit the choice of a catalyst. When the catalyst is water insoluble it will be necessary to separate it from the product by crystallisation in the presence of a solvent for the catalyst, or by distillation. Partially water soluble catalysts can be removed from a water insoluble product by washing. In this respect methyltributylammonium salts are better than the commonly used tetra-n-butyl ammonium analogues.

## 8.4 Solid-liquid reactions

An example is shown in Fig. 6.1. This reaction occurs in solution following dissolution of the potassium fluoride. Tetramethylammonium salts catalyse the process. Under practical conditions due to the low solubility of potassium fluoride ($6 \times 10^{-4}$ M at 120°C) the rate is limited by the dissolution rate; the lifetime of fluoride in solution in 1M substrate at 120°C is 2–3 s. Thus the reaction rate is very sensitive to the surface area of the potassium fluoride. Unlike liquid–liquid reactions, the interfacial area is independent of the agitation regime. This does not mean that agitation is unimportant. It is essential to ensure that the solid is fully dispersed in the reaction mass, otherwise parts of the solution will become depleted in the dissolved solid reactant and the rate will fall accordingly.

## 8.5  Conclusions

As with all aspects of process development, the most important contributions of an understanding of reaction mechanism are to provide new ideas and approaches to the solution of problems, and to ensure that opportunities are not missed because of a failure to comprehend the possibilities.

## Bibliography

Starks C.M., Liotta C. and Halpern M. (1994). *Phase-transfer catalysis: fundamentals, applications and industrial perspectives*, Chapman and Hall.
Danckwerts P.V.(1970). *Gas-liquid reactions*, McGraw-Hill, New York.
L.K. Doraiswamy and M.M. Sharma (1984). *Heterogeneous Reactions: analysis, examples and reactor design; volume 2: fluid-fluid-solid reactions*, John Wiley.
J.H. Atherton (1994). Mechanism in two-phase reaction systems: coupled mass transfer and chemical reaction, in *Research in Chemical Kinetics*, 2, 193.

# 9 Product isolation and work-up

A laboratory multi - stage synthesis will often start from purified materials and include isolation and purification of intermediates at each stage. In contrast, in order to reduce waste for treatment and to contain equipment costs within reasonable limits, an industrial process will have fewer isolation stages and operate with crude intermediates as far as possible. The most popular isolation technique in a laboratory process is multiple crystallisation. At full scale crystallisation is still a very common technique, but by no means always the best. Table 2.4 lists the most common isolation technologies, distillation, liquid extraction, precipitation and crystallisation with comments. In this chapter they are discussed in turn. They are not described in detail, but the basic principles are explained in order to highlight the important issues and provide guidance for process development.

## 9.1   Distillation

Distillation is a very good purification technique, where applicable, but there are often reasons why it cannot be applied. The basic requirements for a good separation are:

1. A significant difference in volatility between the product and the impurities to be separated. A good indicator is difference in boiling point, a difference of the order of 15 $^0$C is required.

2. The distillation temperature and pressure are easily achievable. The upper temperature limit will often be set by the thermal decomposition or potential instability of the mixture. Temperatures in excess of $300^0$C would not normally be encountered in fine chemicals distillation. Pressures are usually between 20 bar pressure and 30 mbar vacuum.

3. The bottom product (heavy) is a liquid at the distillation temperature and pressure. It is possible to operate with a solid or high viscosity bottoms product, but considerably more difficult.

4. The mixture does not form an azeotrope at a composition between the starting  mixture and the purity to be achieved.

An azeotrope is a mixture where the vapour has the same composition as the liquid, therefore on boiling, the composition will not change.

Where the product has a high boiling point, distillation under vacuum or steam stripping are often used and are described in the following sections.

*Basic principles*

The basis for distillation is that when a mixture is boiled, the vapour is richer in the more volatile, light components and the liquid richer in the less volatile heavies. The simplest form of distillation would have the mixture in a still, the vapours condensed and collected and the residue retained in the still. If the difference in volatility between the product and impurities is narrow, fractional distillation will be required. If the vapour is condensed and the resulting liquid reboiled, the vapour will be richer again in lights and so on.

The normal representation is illustrated in Fig. 9.1, for the simple case of a binary mixture, in this example benzene and toluene.

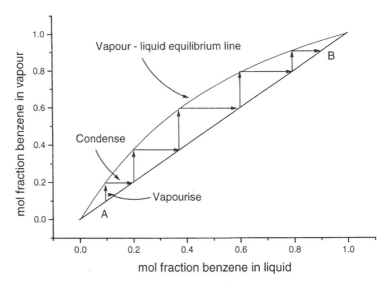

**Fig. 9.1** Benzene - toluene vapour - liquid equilibrium line

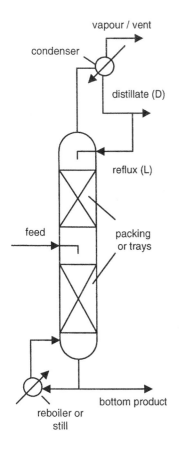

**Fig. 9.2** Schematic typical distillation column layout

The reflux ratio is the ratio of the *molar flow* of liquid returned down the column (*L*) to the molar distillate flow (*D*) as illustrated in Fig. 9.2.

The mole fraction of light component in the vapour (benzene) is plotted against the mole fraction in the liquid. This is the vapour - liquid equilibrium line. The illustration is for a simple ideal case. The steps show the path followed by taking the vapour from point A, condensing it, then vapourising that liquid and collecting the vapour, then condensing, reboiling again until reaching point B, the purified light component.

In practice the consecutive vapourisation and condensation is achieved in a column, illustrated in Fig. 9.2. The vapour passes upwards and the liquid downwards. Industrial columns are usually in the range 0.3 to 10m diameter and 3m up to 75m high. The equilibrium between the liquid and vapour is achieved at each stage throughout the column height by mixing caused by packing or trays. There are many different designs of internals suitable for the whole range of operating conditions, throughput rates and materials. The degree of purification achievable is determined by the number of equivalent equilibrium stages, called theoretical stages, and the amount of condensed overhead vapour returned down the column, the reflux. The height of packing or the number of real trays can be related to the number of theoretical stages required.

The simplest separation is a boil over from a still. This has one stage, no packing and no reflux. Columns for more difficult separations can have many stages, for example 20 to 30 would not be uncommon, and can have a high *reflux ratio*, up to 25.

$$R = \frac{L}{D} \qquad (9.1)$$

Distillations can be batch or continuous. Very large scale petrochemical operations are continuous. Smaller scale production, less than around 1000 - 2000 te an$^{-1}$ tends to be batch. A continuous column will operate at steady state. It will produce top and bottom product streams and have the feed at an intermediate position part way up the column. Intermediate product streams may also be taken off at different column heights when separating a complex mixture.

The focus of this primer is on complex organic chemical manufacture which normally will use batch operation because of the relatively low production rates. A batch distillation will start with the batch charge in the still. The batch will be heated to boiling and the vapour passed up the column to contact with the liquid reflux returning. The top product, called the distillate, is taken off as illustrated. Initially the distillate will be rich in the light components, gradually containing more of the high boilers as the distillation progresses. Similarly the concentration of heavies in the reboiler and the reboiler temperature will rise. There are several ways to operate the column. It can operate at constant distillate purity but reducing distillate rate, by increasing the reflux ratio, or constant rate but accepting increasing higher boilers in the distillate at constant reflux. Intermediate purity products can also be taken.

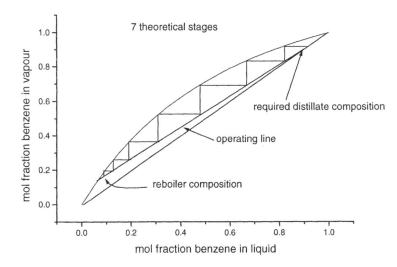

**Fig. 9.3**  Distillation column operating line

Fig. 9.3 shows an operating line on the vapour - liquid equilibrium diagram. This is the relationship between the liquid and vapour compositions at each theoretical equilibrium stage within the column and can be derived by a mass balance.

$$y_n = \frac{R}{R+1}x_{n+1} + \frac{x_D}{R+1} \qquad (9.2)$$

Where $y_n$ is the vapour composition leaving the $n_{th}$ theoretical stage, n is counted upwards with the reboiler being 1; $x_{n+1}$ is the liquid leaving the stage above (Fig. 9.4). $x_D$ is the distillate or top product composition.

The steps between the operating line and the equilibrium curve represent the theoretical stages in the column. In this example the reflux ratio is 15 and there are 7 theoretical stages required to produce a distillate exceeding 90% benzene from a mixture of 10% benzene in toluene. As the batch distillation proceeds, for the example of constant reflux, the operating line retains the same slope, but gradually moves to the left as the benzene content in the reboiler reduces. The benzene content in the distillate will fall. For the case of increasing reflux ratio, the benzene content in the distillate will remain constant, but the slope of the operating line will increase to approach the $45^0$ line as the distillate rate reduces. At this point the column is at total reflux, the reflux ratio is infinite and there is no distillate offtake.

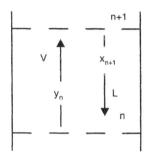

**Fig. 9.4** Liquid and vapour compositions within a column

*Steam stripping*

For organics which are immiscible with water a useful technique to reduce the distillation temperature is to use steam stripping, by injecting steam or water directly into the still. This is an alternative to applying vacuum to reduce the temperature, but works in the same way, by reducing the partial pressure of the organic vapour. In the example, the mixture of 90% toluene, 10% benzene, would boil at $106^0$C at atmospheric pressure, but with steam injected, the mixture will boil at $82.3^0$C. Steam and the benzene rich vapour would be taken over the top, condensed to form two liquid phases and the lighter organic phase separated by gravity.

*Azeotropes*

Fig. 9.5 shows the vapour - liquid equilibrium curve for Acetone - Carbon disulphide. This system has an azeotrope at the composition $x_A$, where the liquid and vapour have the same composition. *A mixture exhibiting azeotropic behaviour cannot be separated by a simple distillation* as the vapour and liquid will reach the same composition and further boiling has no effect. More complex distillation schemes involving other added components or changes in pressure are required.

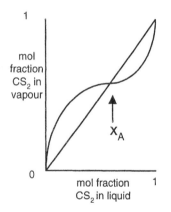

**Fig. 9.5** Acetone - carbon disulphide vapour - liquid equilibrium

*Laboratory practice*

Vapour - liquid equilibrium data are often not available for complex organic molecules and must be estimated, often by methods based on structure and group contribution. As well as the uncertainty in the vapour pressure for the individual compounds, there can be interactions between them which cause deviations from ideal behaviour for the mixture. It is therefore usually necessary to carry out laboratory experiments to determine or confirm the equilibrium data. A simple boil-over test is often sufficient, boiling a sample and collecting samples of the overhead vapour. The liquid in equilibrium in the still can be sampled or the composition calculated by mass balance. For operation under vacuum, it is necessary to set up a sampling system which will allow the distillate to be sampled without affecting the column pressure. *It is also necessary to ensure no reflux, by lagging the apparatus extremely well or even heating the vapour offtake.*

## 9.2   Liquid extraction

Liquid extraction operates by the selective partition of the material to be separated into an immiscible phase which can then be separated, usually by gravity but also possibly by centrifugation. The solvent selection will be based on the equilibrium partition and the ability to separate. The equilibrium partition and the effects of pH, salts and phase transfer catalysts are described in chapter 6.

The ease of separation can be affected by the physical properties and the equipment design, as described in Chapter 7. A number of simple rules of thumb summarised in Tables 9.1 and 9.2 should prove useful:

For a good separation, it is obviously essential for the solvent to be immiscible. Godfrey (1972), Chemtech, June, 359, has provided a simple approach to predicting miscibility from a miscibility number, which can be a useful indicator.

**Table 9.1**   Rules of thumb for liquid contacting for extraction

As a minimum operate with both phases well dispersed (see chapter 7).

Phase inversion should be avoided (see chapter 7).

Safest operation is away from the ambivalent region.

For reproducibility use a standard straight sided vessel e.g 100mm diameter, with baffles and a well designed agitator (see chapter 5).

Check the effect of drop size on rate of extraction by operation at an agitator speed for just dispersed and at twice that speed.

**Table 9.2**   Rules of thumb for separation after extraction

Measure the separation time with either phase dispersed.

Try to remove solids or get them into solution.

Operate at as high temperature as practicable; higher temperature should mean easier separation, but beware increasing mutual solubility.

Remove or avoid the addition of any unnecessary surface active materials.

A solute which transfers readily between the phases could be added to destabilise the interface and aid separation.

*Multi-stage extraction*

For materials which do not partition well, multi-stage extraction can be used. The normal technique for designing a multi-stage extraction unit would either be a column or a series of mixer - settlers, operated semi-continuously. The mixer settler system is more robust and has greater flexibility. Fig. 9.6 is a schematic of a mixer - settler unit with 2  stages. The solvent and feed run countercurrently.

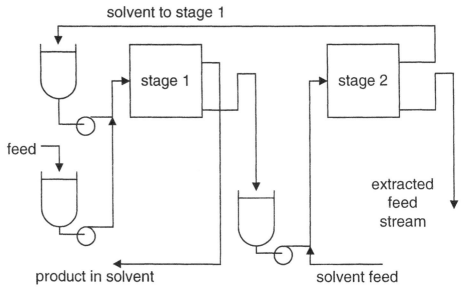

solvent to stage 1

stage 1

stage 2

feed

extracted
feed
stream

product in solvent                    solvent feed

**Fig. 9.6** Countercurrent two stage liquid extraction unit

In a similar way to distillation, it is necessary to calculate the number of equilibrium stages required for the desired degree of extraction. The number of stages ($N_{stages}$) depends upon the equilibrium (partition coefficient $P$) and the ratio of flowrates ($Q_{org}$ for the organic phase and $Q_{aq}$ for the aqueous phase), combined in a separation factor , $M$ :

$$M = P\frac{Q_{org}}{Q_{aq}} \tag{9.3}$$

For $M \neq 1$:

$$N_{stages} = \frac{\ln[\{(x_f - y_s / P) / (x_r - y_s / P)\}(1 - 1/M) + 1/M]}{\ln[M]} \tag{9.4}$$

and for $M = 1$

$$N_{stages} = \frac{(x_f - y_s / P)}{(x_r - y_s / P)} - 1 \tag{9.5}$$

$x_f$ and $x_r$ are the concentrations in the feed and extracted streams. $y_s$ is the concentration in the solvent feed and is usually zero. These equations can be used to estimate the number of stages and ratio of flows required to achieve any desired fraction of the feed left in the extracted stream.

*Settling tests*

The longest time in a batch extraction is usually the time required to separate the mixture afterwards. The separation time is also usually the determining factor for the design of a semi-continuous system. The simplest way to determine the ability to separate is to produce a dispersion of the real materials under well agitated conditions in a tall cylindrical vessel, stop the agitation and observe the movement of the interfaces which appear. Plotting the progress of the clear regions should produce a diagram similar to Fig. 9.7, which is an example for a dispersion of water drops separating from a monochlorobenzene continuous phase. Monochlorobenzene is heavier than water, hence the lower front is produced by water drops settling upwards, leaving clear liquid behind. The upper front is created by water drops coalescing with the interface. The shapes of the coalescing and settling "fronts" are quite distinctive. The slopes of the lines are used to design the large scale unit, by ensuring that the superficial velocity of the dispersed phase at full scale, i.e flowrate /cross sectional area, is lower than the settling velocity determined from the slope of the fronts from the laboratory test.

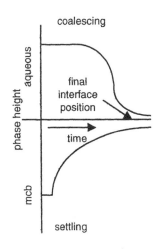

**Fig. 9.7** Separation curves showing the movement of the interface for water drops in a monochloro-benzene continuous phase

## 9.3 Crystallisation and precipitation

Crystallisation is probably the most well known and common laboratory technique for purifying materials. The material can be crystallised from solution or from a melt. For solution it will be driven by causing the concentration in solution to exceed the solubility, either by removing solvent, for example by evaporation, or by reducing the solubility, for example by cooling or by changing the chemical nature of the material to be isolated to a less soluble form, or by changing the nature of the solvent. For melt it will be driven by taking the material to a temperature below the melting point. The objective is to leave the impurities behind in solution or a melt phase and separate off the purified solid.

Usually the pure crystalline solid is separated by filtration or centrifugation, washed and often dried. The objectives of the crystallisation are therefore to produce a solid of the right purity and of a particle size, size distribution and morphology which can be separated and washed easily. Usually this means a narrow size distribution of regular crystals with similar dimensions on all planes and preferably no smaller than 20 to 50 µm (Wood (1997)). *Thin plate-like crystals and very small particles will not separate at all well.*

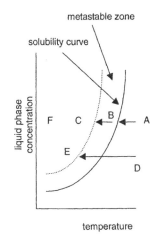

**Fig. 9.8** Metastable zone (Wood, W.M.L.,(1997) Chapter 7 in Chirality in Industry II, Wiley)

### Basic principles

Crystallisation involves two processes, nucleation and growth. Nucleation is the formation of small particles, nuclei, which then grow into larger crystals. The number of nuclei formed determines the number of crystals formed which will grow (primary crystals), thus control of nucleation is essential for control of the crystal size and distribution. The final population of solid particles can contain both primary crystals and crystal fragments generated by attrition, breaking fragments from existing crystals. Therefore control or prevention of attrition is also necessary for control of the particle size.

Nucleation can occur without any crystals present, known as *primary nucleation*, or derived from existing crystals, *secondary nucleation*. Primary nucleation can be homogeneous, caused purely by molecules coming together in the liquid state, or heterogeneous caused by insoluble impurities or foreign bodies even trace quantities, or by a suitable surface. An important concept in controlling nucleation in any crystallisation is the *metastable zone* illustrated in Fig. 9.8. A solution at point A, when cooled will begin to supersaturate at the solubility curve, point B. On continued cooling it will not crystallise, but will increase in supersaturation  up to point C where nucleation begins. Similarly a more dilute solution, point D, will begin to nucleate at a lower temperature, at point E. The curve passing through C and E is the boundary of the metastable zone. This is not a well defined boundary, as the onset of nucleation is very difficult to define and can be rather  unpredictable. However, it is very important to understand that *a supercooled or supersaturated system can exist in a thermodynamically unstable state without crystallising, even for many years.*

It is also important to appreciate the effect of cooling rate on nucleation, hence particle size. A solution rapidly cooled to point F, will have had very little time, if any, for growth, hence will nucleate at F with a very high level of supersaturation. This will result in an extremely high number of nuclei, therefore a very large number of very small crystals, which may produce a high viscosity slurry or be totally unsuitable for separation. Similarly, a large number of nuclei formed by reaching the metastable zone boundary rapidly and continuing to cool can lead to multiple nucleation events, illustrated in Fig. 9.9. Conversely a slow cooling rate will allow the nuclei first formed to grow, hence the concentration in the liquid phase to fall, resulting in a very controlled degree of supersaturation and a small number of large crystals. The ideal cooling profile for controlled growth of the largest particles is to control the degree of supersaturation within the boundary of the metastable zone, Fig. 9.10.

The nucleation rate increases as the undercooling increases, but not without limit. It reaches a maximum and then falls off again, as the viscosity of the system and the mobility of the molecules reduces at low temperatures (Fig. 9.11). It is therefore possible for some materials to form an amorphous glassy state. Glass itself is the most common example, but large organic molecules are also prone to this behaviour.

Secondary nucleation takes place when crystals are present in the system and can occur at much lower supersaturation than is the case for primary nucleation. It can be also more reproducible and leads to the practise of seeding a supersaturated crystallisation mass to induce nucleation and growth. For a mixture of similar molecules, it is possible to induce nucleation of one species, for example one isomer in the presence of another, by seeding, and allow separation by crystallisation.

Crystal growth proceeds by the inclusion of material from the liquid phase into the crystal lattice. For a solution, the rate depends on the area available to grow on, the temperature and the level of supersaturation. There is also an effect of mass transfer resistance around the crystals. For a melt, the driving force is the degree of supercooling. There is also a heat transfer resistance to removing the heat of crystallisation from the surface region. In practise in both systems, growth is generally represented by overall growth rate equations which allow the overall rate of mass deposition per unit area to be calculated.

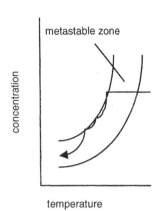

**Fig 9.9** Rapid cooling leading to multiple nucleation events Wood W.M.L., (1997)

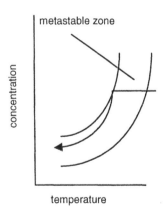

**Fig. 9.10** Ideal cooling profile to control supersaturation

*The effect of impurities*

The very purpose of the crystallisation is to leave the impurities behind in the liquid phase. However, it is not uncommon for some specific impurities to be included in the growing crystal. In particular, *molecules which have the same functional groups and similar shape to parts of the crystallising material can tend to be included.* There is an obvious implication for the ability to achieve the desired purity, if a key impurity is included. A simple laboratory procedure to establish whether this is the case is to take *well washed* crystals, analyse them and establish which impurities are present at high levels relative to others and relative to their concentration in the liquor. It is also possible for relatively insoluble impurities to supersaturate and crystallise in their own right, forming a mixture of crystals rather than being included in a single crystal form. Microscopy should reveal whether there is more than one crystal form present. A knowledge of the relative solubilities is a good indicator. An analysis of the solid phase through time in a batch crystallisation should reveal whether there is a rapid appearance of the impurity, indicating nucleation in its own right and a comparison of the solid composition with liquor should reveal whether it appears upon exceeding a solubility limit.

*Crystal form*

Impurities can also have a very serious effect on crystal form. Fig. 9.12 illustrates the effect of different linear growth rate at different crystal faces on shape. The organisation of the molecules within the crystal lattice causes each face to be dominated by different functional groups. Impurities can therefore interact specifically with one face, inhibiting growth in one plane but not in others. The result can be a change of crystal shape, for example from cubes to plates or needles. In this way a separation which may have been rapid with regular crystals formed from pure material, may become impossible with even trace quantities of growth inhibiting impurities present. A full scale process will be more prone to generating impurities, and therefore the process development must involve a study of the effects of potential impurities on crystal growth.

For further reading on crystallisation see Wood, W.M.L. (1997), and Mullin J.W. (1995)

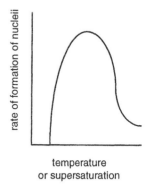

**Fig. 9.11** Nucleation rate vs temperature (Mullin J.W. (1995), Crystallisation, Butterworth-Heinemann

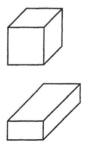

**Fig. 9.12** Different linear growth rate on different faces

# 10   Scale-up

Sharratt P.N., editor (1997).
*Handbook of batch process design,*
Blackie, London.

Scale-up is the process of going from laboratory preparation to whatever scale of manufacture is required to satisfy the market demand. For Fine Chemicals manufacture the scale of operation ranges from a few tonnes up to 10,000 tonnes or more per annum for a large volume agrochemical product. Vessel sizes range from 1 to $50m^3$ or more. Process concentrations vary greatly; ideally no solvent at all is used - this is often possible when one or both of the reactants is a liquid - or else the minimum solvent is used to achieve the desired effect. When a solvent is used, the aim would be to achieve a product concentration of at least 25%, so as to maximise productivity and minimise the costs associated with solvent recovery or disposal.

The science of scale-up is concerned with ensuring that the process performance in terms of yield, quality and operability is not degraded when a process is transferred from laboratory to plant. Scale-up by a factor $10^4$, from 1 litre to 10 $m^3$ or more, is possible provided that the chemistry and the relevant scale-up parameters are understood. The purpose of this chapter is to describe the thought processes and procedures relevant to ensuring successful scale up.

## 10.1   Scale up concepts - why scale-up can be a problem

Scale up is not difficult in principle, but in practice can cause significant problems. The difficulty in practice is that, in order to identify and forestall potential problems, multiple concepts need to be linked in ways which may not be immediately obvious.

Earlier chapters, especially 5, 7 and 8, have shown that the physical properties and characteristics of reaction systems are important in determining how they behave. The larger dimensions and timescales involved in manufacturing processes mean that the parameters determining the performance of the laboratory process cannot all be reproduced at full scale. This means that, for each process which is to be scaled up, it is necessary to identify the physical  factors which will change, and to consider how these will affect both the physical and chemical behaviour at full scale.

For example, if the agitator in a scale model of the plant vessel is operated at the same rpm as full scale, then although the vessel mixing time will be correctly reproduced, the localised power input (which promotes turbulence) will not.

### Some common scale-up difficulties

*Operational (physical) problems*
- Extended reaction times, which can lead to by-product formation.
- Swelling or frothing of the reaction mass, which could lead to expulsion of the reaction mass from the vessel.
- Unexpected rheological characteristics causing difficulties in stirring and heat transfer.

- Slow filtration, most often caused by the particle size being smaller than expected.
- Slow phase separations, leading to extended cycle time.

*Common chemical problems*

- Reduced yield of product
- Increased levels of impurities

## 10.2 Vessel dimensions

Table 10.1 shows the dimensions of typical 10 m$^3$ and 50 m$^3$ vessels used for manufacture, and of a one litre vessel used for laboratory development work. The greater physical dimensions means that heat transfer, gas disengagement rates and phase separations are slower than on the small scale, and that reagent addition times and mixing times within the vessel are longer than those which are typically employed on the small scale. Some of the implications of these changes will be discussed in this chapter.

Increased levels of impurities can adversely influence downstream processing. e.g. phase separations could become more difficult (chapter 7) and the crystal habit of soild products could be altered (chapter 9).

**Table 10.1** Dimensions at some typical vessel sizes

| Vessel working volume, m$^3$ | 50 | 10 | 10$^{-3}$ |
|---|---|---|---|
| liquid depth/m | 5.2 | 2.2 | 0.13 |
| radius/m | 1.75 | 1.2 | 0.05 |

## 10.3 Effect of scale on physical parameters

### Timescales

Several factors combine to ensure that batch processing timescales are extended as the scale is increased. For example, reagent additions take longer; heat transfer is more difficult, because the surface area/volume ratio is much lower; phase separation times are increased because of the greater depth of solution; and filtration times can be longer because filter cake depths are greater.

To fill a 10 m$^3$ vessel via a 5 cm diameter pipe would typically take 2 hr.

### Heat transfer

Scale-up from 1 litre to 10 m$^3$ involves a reduction in the surface area to volume ratio of 22.5 for geometrically equivalent vessels. Heat transfer rates are correspondingly reduced. For equipment using a similar heating or cooling fluid at the same temperature difference between reaction mass and heat transfer fluid, this means that reactions which are heat transfer limited will take many times longer on the full scale. Implications for reactant and product stability over the extended processing times need to be considered. The need to speed up heat transfer may lead to the use of greater extremes of temperature at heat transfer surfaces, which could cause decomposition of materials being processed.

The calculation assumes cylindrical vessels with a flat bottoms; dimensions as in Table 10.1

Processes in which the overall processing time is determined by the rate of supply or removal of heat are described as 'heat transfer limited.'

A further cause for concern is the possibility of exothermic reactions becoming uncontrollable because of the inability to remove heat at a sufficient rate. An accumulation of unreacted starting materials can become extremely hazardous if, when reaction does initiate, it cannot be controlled. This could occur because the reaction temperature is initially too low, or because of some chemical inhibition of a catalysed process. These considerations are a major factor in favour of the 'semi-batch' operating conditions which are so widely used. Avoidance of dangerous exotherms is achieved by the controlled addition of one or more reactants, with some means of checking that reaction is occurring as the addition proceeds. This may not have to be a chemical analysis; monitoring the heat output may be an adequate method of ensuring safety.

**Two-phase systems**

This topic has been covered in Chapter 7. Problems are exacerbated at the large scale because specific power inputs are usually less than on the laboratory scale. Reactor configuration is crucial. An appropriate agitator and baffling must be used (Chapter 7). Problems can easily arise when ad-hoc equipment is used which is inappropriate for the required duty.

> For example, an 'anchor' agitator is almost always an inappropriate choice where two-phase systems are involved, because it will not generate adequate interfacial area.

*Suspension of solids*

As a minimum, in order to achieve reproducibility, there should be no solid permanently residing on the vessel base. Agitation above that necessary to achieve this is unlikely to be beneficial.

*Contacting of immiscible liquids*

To achieve reproducibility the process should be designed to give a similar interfacial area (drop size) on laboratory and full scale. Reasons for this have been discussed in Chapter 8, and the means of designing for this have been outlined in Chapter 7. It may be important to control which phase is dispersed, and to avoid phase inversions (Chapter 7).

*Phase separation*

In an unstirred vessel phase separation times are roughly proportional to the liquid depth, which may well be up to 50 times larger at full scale *vs* the laboratory. In laboratory work phase disengagement times of less than a minute should be sought. Again, attention to phase continuity is essential. Small quantities of surface active agents can have a major influence on phase separation rates, by inhibiting droplet coalescence.

*Gas disengagement*

For a reaction which generates gas, the rate of formation of gas is proportional to volume, but the rate of gas disengagement is proportional to surface area. Since the surface area : volume ratio decreases with scale, the proportion of gas held up as bubbles in the liquid phase increases. If this factor is not properly considered then the swell may be sufficient to expel some of the batch from the vessel. The scale up parameter is the superficial velocity of the gas phase. An escape superficial velocity of about $0.1$ m s$^{-1}$ is the maximum practicable, above which serious swelling of the liquid will occur, and a two-phase liquid vapour mixture may be carried into the upperworks of the vessel. If surface active materials are present, the

> Superficial gas velocity is the mean gas velocity above the solution, in other words, the volumetric gas flow rate divided by the cross sectional area of the vessel.

maximum practicable superficial velocity could be much less, due to frothing of the liquid.

A common situation is the neutralisation of acids with bicarbonates. Generation of $CO_2$ may not be a problem on the small scale, but can become so as the scale is increased, because the surface area : volume ratio is much less at manufacturing scale, so that the gas cannot escape at the same rate.

**Rheology**

For Newtonian fluids, scale up does not introduce any problems derived from rheology. For other types of fluid, particularly those which show shear thinning, difficulties can arise because of the wide variation in shear rate in a large vessel. Such a fluid may effectively set solid in low shear regions away from the agitator, leading to heat and mass transport problems. Unexpected increases in viscosity can lead to swelling of the vessel contents by reducing gas disengagement rates.

See chapter 5

**Distillation**

Distillations at the full scale tend to be more difficult than on the small scale. One reason can be that it is difficult to eliminate reflux on the laboratory scale even in a simple Claisen head, and thus there may be two or three theoretical plates of separation where none was intended. An over-optimistic view of the ease of separation may be obtained. To avoid this the uplift should be heated to above the boiling point of the distillate.

Distillation rates are usually limited by the ability to input heat to the vessel contents. Because of the reduced surface area to volume ratio on the full scale, distillations can therefore take much longer. This may lead to chemical selectivity problems if reactants are unstable.

## 10.4 Chemical selectivity issues

Factors which commonly influence chemical selectivity include the following:

- Processing times, which are frequently extended on the large scale.
- Mixing time
- Interfacial area in two-phase reactions

Often problems are caused by interaction of the chemistry of the process with physical variables. A useful principle to remember where a chemical stage is involved is that

*chemical rate constants are scale independent, whereas most related physical processes are not.*

This is the primary cause of the failure to identify scale-up problems associated with chemical selectivity. An understanding of the nature of any alternative reaction pathways is essential to achieving secure scale-up. Several examples of competing reaction pathways are given at the beginning of Chapter 4.

Hoyle W., editor (1998). *Pilot plant and scale-up of chemical processes I J* Royal Society of Chemistry, London.

Prior to scale-up processes should always be trialled in the laboratory using the same timescales as will be enforced by plant requirements.

A well known laboratory example is the preparation of acetaldehyde by oxidation of ethanol, where the product (b.p. 21°C) is removed as it is formed.

For example, when a mono-nitration product is required in a process where dinitration is rapid, the order of addition would always be nitric acid into substrate. This should trigger a warning that dinitration *coul d*occur in mixing transients (See Chapter 5).

Agitator speeds at full scale are in the range 30–120 rpm. These appear unrealistically low on the laboratory scale.

## Timescales

Probably the commonest cause of reduced selectivity on scale up is the instability of either reactants or product over the longer timescale needed for processing on the full scale. This factor is easily checked, so that appropriate action can be taken. If *reactants* are unstable under processing conditions then the usual solution is to add those which are unstable to the reaction mass under conditions chosen to maximise the selectivity. Chapter 4 gives some examples of aspects to consider. More difficult are the cases where the *product or an intermediate* is unstable. In extreme cases the reactor configuration may be dictated by the need for a short residence time. Again, Chapter 4 gives examples. It may be possible to remove a product continuously by distillation.

## Mixing effects in pseudo-homogeneous systems

This topic has been discussed in Chapter 5. Because power inputs per unit volume are much less on the large scale, and vessel mixing times longer, mixing effects are more pronounced. Neutralisation of an acid catalyst in an ester at the end of an esterification process is discussed in Chapter 5.

It is always useful to ask the question 'what would happen if the order of addition of reagents were reversed?' Although there may be no intention to do this, any serious negative concern generated should lead to a more detailed consideration of possible mixing problems.

A useful empirical test to determine whether macromixing effects will impair selectivity on scale up is to carry out the process in a geometrically identical model of the equipment, with the agitator running at the same speed as the full scale, and the reagents added over the same timescale. This will provide a bulk circulation time approximately equal to full scale, and will show up any problems due to that particular parameter. Other factors will not be mimicked e.g. suspension of solids or dispersion of two-phase systems, so this test must be used with care.

### Two phase reactions

Two phase reactions are particularly difficult to scale up because the number of factors which need to be controlled is considerably greater than for single phase systems. The most important scale-up parameter is the interfacial area. All two-phase reactions show a rate dependence on interfacial area; whether or not the chemical selectivity is affected depends on how the rates of competing reactions are influenced by mass transfer rates. In order to ensure satisfactory scale up it is necessary to identify any competing chemistry and then to consider how this chemistry may be affected by changes in mass transfer rates. Some examples are given in Chapter 8.

## 10.5   Summary

It has been the intention of this brief Chapter to show in qualitative terms the principles involved in scale up from laboratory to full scale manufacture. It is an area where there are great synergies to be gained from interdisciplinary working between chemists and chemical engineers.

# Index